枫泾镇 Fengjing Town

枫泾镇 Fengjing Town

枫泾镇 Fengjing Town

建筑可阅读书系

文物视角中的江南：
上海古镇
编著：上海市文物保护研究中心

编委会

主任：方世忠
副主任：向义海
编委：李晶、邓军、舒晟岚、翟杨

中文文字：高文虹、蒋薇、李弥
英文文字：栾建红
英文校对：葛文荟、李弥
摄影：许一凡，等
绘图：苏诗淼、岳欣芷

鸣谢（排名不分先后）：
上海市浦东新区文物保护管理所
上海市嘉定博物馆
上海市金山区博物馆
上海市青浦区博物馆
上海淞沪抗战纪念馆（上海市宝山区文物保护管理所）
上海市宝山区罗店镇社会事业发展服务中心
上海市金山区张堰镇文化体育服务中心
上海市浦东新区新场镇文化服务中心
上海市青浦区朱家角镇社区文化活动中心

"Stories of Shanghai Architecture" Series

Jiangnan from the Perspective of Cultural Relics: Ancient Towns in Shanghai
Shanghai Cultural Heritage Conservation and Research Center

Editorial Board

Director: Fang Shizhong
Deputy director: Xiang Yihai
Editorial members: Li Jing, Deng Jun, Shu Shenglan, Zhai Yang

Chinese text: Gao Wenhong, Jiang Wei, Li Mi
English text: Luan Jianhong
English proofreading: Ge Wenhui, Li Mi
Photography: Xu Yifan, et al.
Drawing: Su Shimiao, Yue Xinzhi

Acknowledgements (in no particular order):
Shanghai Pudong New Area Cultural Relics Protection and Management Institute
Shanghai Jiading Museum
Shanghai Jinshan District Museum
Shanghai Qingpu District Museum
Shanghai Songhu Memorial Hall for the War of Resistance Against Japanese Aggression (Shanghai Baoshan District Heritage Conservation and Administration Institute)
Shanghai Baoshan District Luodian Town Social Development Service Center
Shanghai Jinshan District Zhangyan Town Culture and Sports Service Center
Shanghai Pudong New Area Xinchang Town Cultural Service Center
Shanghai Qingpu District Zhujiajiao Town Community Cultural Activity Center

Jiangnan from the Perspective of Cultural Relics
Ancient Towns in Shanghai

文物视角中的江南
上海古镇

上海市文物保护研究中心 编著
Shanghai Cultural Heritage Conservation and Research Center

同济大学出版社·上海
TONGJI UNIVERSITY PRESS · SHANGHAI

前言

　　江南文化是上海三大文化品牌之一，意蕴深厚、韵味悠长。从枫泾、新场、朱家角等历史文化名镇，到唐经幢、真如寺大殿、泖塔等优秀古代建筑，从豫园、秋霞圃、醉白池的曲径通幽，再到桥涵码头、传统民居的市井生活，造就了上海独特的江南文化特质。建筑可阅读书系之文物视角中的江南系列，以文物建筑为主要视角，从古镇、古园林、古桥、古塔、寺庙祠堂、传统民居六方面对上海江南文化进行诠释。作为系列丛书的开篇，本书从集江南文化元素之大成的古镇切入，探寻上海从秀丽江南向繁华都市演进的过程。

　　上海地处江南，紧邻东海，因水而生，在上万年前本是汪洋大海。6000多年前，上海地区的古海岸线"冈身"成形，纵贯现嘉定、青浦、松江、闵行、金山五区，其以西地区开始有了人类活动的痕迹。秦、西汉时期，上海地区属会稽郡，东汉时期，松江地区的华亭首次作为地名在史书中出现，境内的农田、水利得到了一定的开发。唐代，位于吴淞江出海口的青龙港兴起，吴淞江以南的松江府城区域和以北的嘉定县境域日益繁荣，今浦东、奉贤地区则逐步成陆，发展出盐田和官方驻兵屯守的聚居点。在这些地方，以河流为交通动脉，因商品贸易形成草市，商店、民居傍水而立，逐步发展为传统市镇。

　　明清时期，上海地区市镇基本成形，大部分市镇从事棉纺织业和米粮业，并成为区域的贸易中心，如朱家角、枫泾等；另有浦东市镇依托滨海的地理位置发展制盐业、捕鱼业和耕织业，如新场、下沙等；还有少部分沿海市镇由明代的卫、所、堡、墩演化而来，如川沙、金山卫等。这些市镇的地理环境和经济形态塑造了其空间形态、街巷格局和建筑风貌，商品贸易的往来促进了镇内长街的成形，发达的水系促成了古桥、码头、驳岸等建筑的修筑。镇内建筑与河道水系和田园树林自然融合，构成了江南市镇丰富错落的肌理。民居建筑基本呈现了江南民居粉墙黛瓦的风格，又因人口迁入和贸易往来混入了徽派马头墙和屏风墙等外来建筑元素，展现了上海"海纳百川"的地域特征，开埠后更是引入了西方彩色玻璃、马赛克地砖等建筑材料，呈现出了"中西融合"的文化特点。

在物质文明形态塑造的同时，上海地区的文学、艺术等精神文明形态也在不断发展。陆机、陆云等文人诞于华亭，苏轼、米芾等名家客居青龙，以董其昌为首的松江画派统治了明晚期的画坛，其开阔、淡泊的笔墨意境深刻地影响了中国文人画的风格和走向。此外，上海地区文化形态也深刻反映在节庆、工艺、饮食、服饰等非物质文化遗产中。青浦田山歌、嘉定竹刻、徐行草编、南翔小笼、高桥绒绣、罗店划龙船等技艺不仅反映了上海地区先民的生活方式、思维观念和审美情趣，也展现了上海市镇以农耕文明为主，兼具海洋文明的文化特征。

如今，上海地区仍保留有一定数量的传统古镇，主要分布在金山、青浦、浦东、嘉定、宝山等地。其中历史价值丰富、保存状况较好的 11 处——枫泾镇、朱家角镇、新场镇、嘉定镇、南翔镇、练塘镇、张堰镇、高桥镇、金泽镇、川沙新镇及罗店镇，于 2015—2019 年先后被公布为中国历史文化名镇。这些古镇代表了上海的历史文脉，告诉了我们上海从何而来。它们也是构成上海城市品格不可缺少的部分，是上海去向何处的基石，对城市精神的塑造和软实力的提升有着重要的作用。因此，保护好上海的水乡古镇，将上海的"江南文化"阐释好、传承好、发扬好，是社会各界需要思考和努力的方向，而"建筑可阅读"就是一种重要的方式和途径。

近几年，为深入贯彻落实"人民城市人民建，人民城市为人民"重要理念，上海市委、市政府全面推进"建筑可阅读"工作，让人们通过阅读建筑的形式走近城市历史、触摸城市的文化印记。在此背景下，上海市文物保护研究中心在上海市文化和旅游局（上海市文物局）的指导下推出本书系的首册《文物视角中的江南：上海古镇》，对上海 11 处中国历史文化名镇进行解读，诠释"建筑是可阅读的，街区是适合漫步的，城市始终是有温度的"城市内涵。本书对这些古镇的历史渊源、文化内涵和价值特色进行梳理，对古镇中的文物点位进行介绍，结合照片等展示方式，力求在历史、文化、空间等多维度中客观呈现"江南文化"。为兼具专业性、实用性与趣味性，本书配有交通方式、旅行线路、美食推荐和英文翻译，可满足国内外游客深度阅读建筑的需求。希望读者通过这本书，走进上海的江南古镇，聆听上海的江南往事，感受这座城市的魅力。

Foreword

Jiangnan (Jiannan refers to the region located south of the Yangtze River in China) Culture is one of the three major cultural brands in Shanghai, with profound connotations and timeless charm. From famous historical and cultural towns like Fengjing, Xinchang, and Zhujiajiao, to outstanding ancient architecture such as Tangjingzhuang (Tangjingzhuang refers to a type of commemorative monument inscripted with Buddhist scriptures and teachings, serving to commemorate the spread and translation of Buddhism during the Tang Dynasty), Zhenru Temple Hall, and Mao Tower and from the secluded paths of Yu Garden, Qiuxia Garden and Zuibai Pond, to the bustling life of bridge and wharf terminals and traditional residences, Shanghai has nurtured its unique Jiangnan cultural characteristics. The "Jiangnan from the Perspective of Cultural Relics" volumes of "Stories of Shanghai Architecture" series, with cultural relics and architecture as the main perspective, interprets Shanghai's Jiangnan culture from six aspects: ancient towns, ancient gardens, ancient bridges, ancient pagodas, ancient temples and halls, and traditional residences. As the opening book of this series, this book starts from the ancient town that embodies the essence of Jiangnan culture, exploring Shanghai's evolution from a beautiful Jiangnan region to a bustling metropolis.

Shanghai is located in the Jiangnan region, adjacent to the East China Sea. It originated from water and was originally a vast ocean thousand of years ago. More than six thousand years ago, the ancient coastline of Shanghai, known as "Gangshen", took shape, spanning the present-day districts of Jiading, Qingpu, Songjiang, Minhang, and Jinshan. Human activities began to leave traces in the western part of the region. During the Qin and Western Han dynasties, Shanghai belonged to Kuaiji Prefecture. During the Eastern Han Dynasty, the name "Huating" in Songjiang area first appeared in historical records, and the cultivation of farmland and water conservancy in the region began to develop. In the Tang Dynasty, Qinglong Port at the mouth of the Wusong River prospered, and the urban area of Songjiang Prefecture south of the Wusong River and the Jiading County north of it became increasingly prosperous. The present-day Pudong and Fengxian areas gradually became land, developing salt fields and military garrisons. In these areas, rivers served as the main transportation routes, and grass markets formed due to commodity trade. Shops and residences were built along the waterfront, gradually developing into traditional market towns.

During the Ming and Qing dynasties, the market towns in the Shanghai area took shape.

Most of these market towns were engaged in cotton textile and rice trade, making them regional trade centers, such as Zhujiajiao and Fengjing. In addition, Pudong market towns developed salt-making, fishing, agriculture and textile industries due to their coastal location, such as Xinchang and Xiasha. There were also a few coastal market towns that evolved from the Ming Dynasty's defensive system of Weis, Suos, Baos and Duns ("Weis" were military stations or forts strategically located along the coast, serving as the first line of defense against enemy forces; "Suos" were smaller garrisons or guard posts that were usually situated close to Weis, acting as support units; "Baos" were fortified strongholds that served as important defense points, often constructed at key locations along the coast or on islands; "Duns" were small earthworks or low defensive walls built to reinforce the defense system, typically located in strategic areas such as hills or narrow points, helping to secure the coastal region), such as Chuansha and Jinshanwei. The geographical environment and economic structure of these market towns shaped their spatial layouts, street patterns, and architectural styles. The trade activities led to the formation of long streets in the towns, while the developed water systems resulted in the construction of ancient bridges, docks, and embankments. The integration of town architecture with rivers, water systems, and fields created a diverse and intricate texture of Jiangnan market towns. The residential buildings mainly exhibited the typical style of Jiangnan houses with whitewashed walls and grey tiles. Then, due to population migration and trade activities, external architectural elements like Huizhou Horse Head Wall and folding screen walls were also incorporated, showcasing the regional characteristics of Shanghai's open, inclusive, and accepting of diversity to "embrace all the rivers". After the opening of the port, Western building materials such as colored glass and mosaic tiles were also introduced and adopted, presenting a cultural characteristic of "fusion of the East and West".

While material civilization was shaping in the Shanghai area, the forms of spiritual civilization such as literature, art were also continuously developing. Literary figures like Lu Ji and Lu Yun were born in Huating. Great masters like Su Shi and Mi Fu once resided in Qinglong. The Songjiang Painting School, led by Dong Qichang, dominated the painting world in the late Ming period with its open and detached artistic style, profoundly influencing the style and direction of Chinese literati painting. In addition, cultural forms in the Shanghai area were also deeply reflected in intangible cultural heritages such as festivals, crafts, cuisine, and clothing. The farm song of Qingpu, the bamboo carving of Jiading, the straw weaving of Xuhang, the Xiaolong (small steamed buns) of Nanxiang, the woolen embroidery of Gaoqiao, and the dragon boating customs of Luodian not only reflected the lifestyle, mindset, and aesthetic taste of the ancestors in the Shanghai area, but also showcased the cultural characteristics of Shanghai market towns, which were primarily influenced by agrarian civilization while incorporating elements of maritime civilization.

Nowadays, the Shanghai area still retains a certain number of traditional ancient towns, mainly distributed in Jinshan, Qingpu, , Pudong, Jiading, Baoshan, and other places. Among them, 11 sites with rich historical value and well-preserved conditions, including Fengjing Town, Zhujiajiao Town, Xinchang Town, Jiading Town, Nanxiang Town, Liantang Town, Zhangyan Town, Gaoqiao Town, Jinze Town, Chuansha New Town, and Luodian Town, have been declared as national famous historical and cultural towns successively from 2015 to 2019. These ancient towns represent the historical context of Shanghai and tell us where Shanghai comes from. They are also an essential part of Shanghai's urban character and the foundation for its future. They play an important role in shaping the city's spirit and enhancing its soft power. Therefore, protecting the water towns in Shanghai, interpreting, inheriting, and promoting the "Jiangnan culture" of Shanghai is a direction that needs to be considered and worked on by all sectors of society. The concept of "Stories of Shanghai Architecture" is an important way and means to achieve this.

In recent years, in order to fully implement the important concept of "People's City built by the People, People's City made for the People", the Shanghai Municipal Party Committee and the Municipal Government have comprehensively promoted the "Stories of Shanghai Architecture" work, which enables people to explore the city's history and touch its cultural imprints through reading architectures. In this context, under the guidance of the Shanghai Municipal Administration of Culture and Tourism (Shanghai Municipal Administration of Cultural Heritage), the Shanghai Cultural Heritage Conservation and Research Center has launched *Jiangnan from the Perspective of Cultural Relics: Ancient Towns in Shanghai*" as the first volume of this series, to explore the 11 national famous historical and cultural towns in Shanghai and explain the city's connotation: "Architecture is readable, neighborhoods are walkable, the city is amiable". This book compiles the historical origin, cultural connotation and value of these ancient towns, introduces the cultural relics in the ancient towns, and combines photos and drawings to objectively present the "Jiangnan culture" in multiple dimensions such as history, culture and space. In order to combine professionalism, practicality and fun, the book is equipped with transportation guide, tourist routes, local delicacies and English translation, which can meet the needs of domestic and foreign tourists to read the architecture in depth. It is hoped that readers can step into the Jiangnan ancient towns of Shanghai through this book, listen to the stories of Shanghai's past, and experience the charm of this city.

目录

Contents

上海古镇区位示意图

Location Diagram of Shanghai Ancient Town

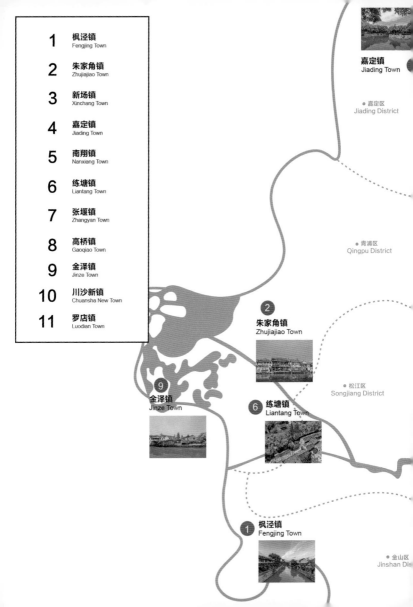

嘉定镇
Jiading Town

● 嘉定区
Jiading District

● 青浦区
Qingpu District

2 朱家角镇
Zhujiajiao Town

● 松江区
Songjiang District

9 金泽镇
Jinze Town

6 练塘镇
Liantang Town

1 枫泾镇
Fengjing Town

● 金山区
Jinshan Dis

11 罗店镇
Luodian Town

静安区
Jing'an District

宝山区
Baoshan District

杨浦区
Yangpu District

8 高桥镇
Gaoqiao Town

5 南翔镇
Nanxiang Town

普陀区
Putuo District

虹口区
Hongkou District

长宁区
Changning District

10 川沙新镇
Chuansha New Town

黄浦区
Huangpu District

徐汇区
Xuhui District

浦东新区
Pudong New Area

闵行区
Minhang District

3 新场镇
Xinchang Town

奉贤区
Fengxian District

张堰镇
Zhangyan Town

本图为位置示意，与实际尺寸不符
Illustration is not proportional to the actual scale

枫泾镇   Fengjing Town

1   枫泾消防纪念塔及东区火政会
枫泾镇枫思路 51 号北侧；生产街 122 号、
123 号、124 号

1   *Fengjing Fire Fighting Memorial Tower and East District Fire Station*
North side of No.51, Fengsi Road; No.122, 123, 124, Shengchan Street , Fengjing Town

2   人民公社旧址
枫泾镇和平街 85 号

2   *Former Site of the People's Commune*
No.85, Heping Street, Fengjing Town

3   朱学范故居
枫泾镇新泾路 200 号

3   *Zhu Xuefan Former Residence*
No.200, Xinjing Road, Fengjing Town

4   程十发祖居
枫泾镇和平街 151 号

4   *Cheng Shifa Ancestral Residence*
No.151, Heping Street, Fengjing Town

5   蔡以台宅
枫泾镇青枫街 34 号、35 号、36 号

5   *Cai Yitai Residence*
No.34, 35, 36, Qingfeng Street, Fengjing Town

6   郁家祠堂
枫泾镇新泾路 276 号

6   *Yu Family Ancestral Hall*
No.276, Xinjing Road, Fengjing Town

7   致和桥
枫泾镇南大街 88 号东侧

7   *Zhihe Bridge*
East side of No.88, South Street, Fengjing Town

8   宝源桥
枫泾镇下塘街 25 号北侧

8   *Baoyuan Bridge*
North side of No.25, Xiatang Street, Fengjing Town

图例 LEGENDS

全国重点文物保护单位
National priority protected site

上海市文物保护单位
Shanghai city-level protected site

区级文物保护单位
District-level protected site

区文物保护点
District-level protected place

其他景点
Other attractions

历史风貌区范围
Scope of historical disctrict

游览路线
Touring route

枫阳路 Fengyang Road

枫泾北大街 Fengjing North Street

白牛塘 Baiu River

枫杰路 Fengjie Road

枫泾消防纪念塔
Fenging Fire Fighting
Memorial Tower

施王桥
Shiwang Bridge

三百园
Sanbai Garden

清风桥
Qingfeng Bridge

人民公社旧址
Former Site of
the People's Commune

郁家祠堂
Yu Family
Ancestral Hall

朱学范故居
Zhu Xuefan
Former
Residence

竹行桥
Zhuxing Bridge

泰平桥
Taiping Bridge

古戏台
Ancient Theater

枫泾市河 Fengjing Shiha River

程十发祖居
Cheng Shifa
Ancestral Residence

致和桥
Zhihe Bridge

东区火政会
East District
Fire Station

宝源桥
Baoyuan Bridge

蔡以台宅
Cai Yitai Residence

亭枫公路 Tingfeng Highway

朱枫公路 Zhufeng Highway

本图为位置示意，与实际尺寸不符
illustration is not proportional to the actual scale

枫泾镇
Fengjing Town

交通指南：
地铁 1 号线至莲花路站，转乘公交莲枫专线至枫阳路北大街站；
地铁 1 号线至锦江乐园站，转乘公交枫梅线至枫泾牌楼站。

Transportation Guide:
Metro Line 1 to Lianhua Road Station, transfer to Bus Lianfeng Line to Fengyang Road North Street Station;
Metro Line 1 to Jinjiang Park Station, transfer to Bus Fengmei Line to Fengjing Pailou Station.

吴根越角，千年枫溪

枫泾镇位于上海市西南，地处沪浙交会，水陆交通便利，是通往西南各省的要冲门户。这里既是天宝物华的江南水乡，也是明清以来贸易繁盛的商业重镇，清代文人程兼善的《枫溪棹歌》称"楼阁千家半傍河，露台风月晚来多。阿侬旧住枫溪曲，爱向烟波唱棹歌"，道出了枫泾别样的江南风情。

枫泾历史悠久，2000多年前便有先民在这片土地上繁衍生息。南朝梁天监元年（502）建仁济道院。唐宋以来，这里寺院道观遍布，人烟渐多。元至元十二年（1275）正式建镇，名为白牛镇，明末改称枫泾镇，另有清风泾、芙蓉镇的雅称。明清两代，枫泾商贾汇集，市井兴旺，为华亭县西部繁华之地。所产"枫泾布"，质地牢固，价廉物美，闻名江南数省，素有"买不完枫泾布，收不尽魏塘纱"之誉。清咸丰十年（1860），清兵和太平军于此地激战，大批房屋被焚毁。抗日战争期间，日机轰炸，很多房舍化为瓦砾，市镇一时衰落。

由于地处古时吴越交会，枫泾明代开始以镇中的界河为界，南北分治，南镇属嘉兴府嘉善县，北镇属松江府华亭县，这种分界持续了明清两个朝代。"界桥两岸分南北，半隶菁城半魏塘"，实为分治印证。直至近代，1951年3月南镇并入北镇，合并归松江县管辖。1966年10月，枫泾镇由松江县划入金山县，隶属金山区。

受南北分治影响，古镇形成了特有的两镇双核心格局，虽历经战祸，但空间格局保存完好，风貌特征明显。现镇域内水网遍布，河道纵横，古有"三步两座桥，一望十条港"之称。生产街、南大街、北大街等街巷，整体风貌保存较好。街巷两侧建筑鳞次栉比，多为传统的砖木结构，大多维持传统的前店后宅形式。两镇在百年发展过程中，互为补充、和谐融洽，是"吴根越角"的象征。

枫泾人文荟萃、人杰地灵，历史上有顾氏、许氏、沈氏、郁氏等众多名门望族，艺术文化繁荣。现在的枫泾不仅是享誉海内外的金山农民画发源地之

一，而且这里的蓝印花布、家具雕刻、灶壁画、花灯、剪纸、绣花、编织等民间艺术也是源远流长，更有枫泾丁蹄制作技艺、上海黄酒传统酿造技艺、敛痔散制作技艺、金山故事等非遗项目，文化资源十分丰富。

近年来，枫泾镇多举措并行，倡导文旅融合，打造非遗新体验，举办了"古镇非遗购物节""古镇文化街非遗市集"等主题活动，赋予古镇更加深厚的文化内涵。行走枫泾，可漫步廊桥，煮酒品茗，重拾江南水乡的闲适生活。

不可移动文物资源

2005年9月，枫泾镇被公布为第二批中国历史文化名镇，其中历史文化风貌保护区面积104.1公顷，核心保护区面积31.24公顷。镇域内现有不可移动文物44处，其中上海市文物保护单位3处，区级文物保护单位7处，区文物保护点34处。

A town rooted in Wu and Yue, with a thousand-year Maple Creek

Fengjing Town is located in the southwest of Shanghai, at the intersection of Shanghai and Zhejiang, with convenient water and land transportation, serving as the key gateway to the southwest provinces. It is not only a water town of Jiangnan with heavenly treasures, but also a commercial town with prosperous trade since the Ming and Qing dynasties. Cheng Jianshan, literati of the Qing Dynasty, wrote "Fengxi Rowing Song" that said "Thousands of buildings and pavilions are half beside the river, and there are many winds and moons on the terrace in the evening. Anon ('I' in Shanghai dialect) used to live in Fengxi, and loved to sing rowing song to the misty waves", beautifully depicting the unique Jiangnan flavor of Fengjing.

Fengjing has a long history. There were ancestors living on this land more than 2000 years ago. In the 1st year of the Tianjian reign of Liang in the Southern Dynasty (502), the Renji Taoist Temple was built. The areas have been dotted with temples and Taoist temples from the Tang and Song dynasties, and have become increasingly populated. In the 12th year of the Zhiyuan reign of the Yuan Dynasty (1275), the township was formally established as Bainiu, and was renamed Fengjing at the end of the Ming Dynasty, with elegant names such as Qingfeng (Breeze) Jing and Furong (Lotus) Town. During the Ming and Qing dynasties, Fengjing was a place where the merchants gathered and markets flourished in the western part of Huating County. Fengjing cloth produced there was famous among Jiangnan provinces for its firm texture and inexpensive price, has gained the reputation of "you can never buy enough Fengjing Cloth or collect enough Weitang Yarn". In the 10th year of the Xianfeng reign of the Qing Dynasty (1860), the Qing army and the

Taiping army fought a fierce battle here, resulting in a large number of houses being burned down. During the War of Resistance Against Japanese Aggression, many houses were bombed into ruins by Japanese air force. From then on, The town was in decline for a while.

Due to its location at the crossroads of the ancient states of Wu and Yue, Fengjing has been divided into north and south since the Ming Dynasty, with the Jiehe River in the town as the dividing line. The southern part of the town belonged to Jiashan County of Jiaxing Prefecture while the northern part belonged to Huating County of Songjiang Prefecture. This division lasted through the Ming and Qing dynasties. "Boundary bridge divides the town into north and south, half belonging to Rongcheng while half to Weitang", is a proof of the partition. Until modern times, in March 1951, the south town was merged with the north town to be under the jurisdiction of Songjiang County, and in October 1966, Fengjing Town was transferred from Songjiang County to Jinshan County, belonging to Jinshan District.

Affected by the partition, the ancient town has formed a unique pattern of two cores in two towns. Although it has experienced many wars, the spatial pattern of the ancient town is well-preserved with obvious landscape features. Now the town area is densely covered with water networks and rivers, which has been known as "three steps to meet two bridges, one glance to see ten harbors" before. The overall style of many streets such as Shengchan Street, South Street, North Street is well preserved. Buildings lined up on both sides of the streets are mostly traditional brick and wood structure, maintaining the traditional form of front stores and back houses. The two towns have complemented each other and harmonized in the process of development for a hundred years, symbolizing the "Wugen Yuejiao" (its root in the ancient states of Wu and Yue).

Fengjing is a place abundant with cultural heritage and talented people. Historically, there were many famous families such as Gu, Xu, Shen and Yu. Art and culture flourished here. Nowadays, Fengjing is not only one of the birthplaces of Jinshan Farmer's Painting, which is famous both at home and abroad, but also has a long history of folk arts such as blue prints, furniture carvings, zao (kitchen stove) mural paintings, lanterns, paper-cutting, embroidery, weaving and so on. There are also intangible cultural heritage projects such as Fengjing Ding hooves (special "Braised Pig's Feet" by Ding

枫泾镇参观指南 Fengjing Town Visiting Guide

游览路线：
清风桥→竹行桥→东区火政会→泰平桥→程十发祖居→古戏台→人民公社旧址→三百园→施王桥

Tourist Route:
Qingfeng Bridge → Zhuxing Bridge → East District Fire Station → Taiping Bridge → Cheng Shifa Ancestral Residence → Ancient Theater → Former Site of the People's Commune → Sanbai Garden → Shihwang Bridge

古镇美食特产：
丁蹄、石库门黄酒、阿六烧卖、天香豆腐干、状元糕

Local Specialties:
Ding Hooves, Shikumen Yellow Wine, Ah Liu Shaomai, Tianxiang dried Tofu, Zhuangyuan Cake

Yixing Deli) cooking technique, Shanghai yellow wine traditional brewing technique, *lian zhi san* (medicines for hemorrhoids made from Chinese herbs) production technique, Jinshan stories, etc. The cultural resources in Fengjing are extremely rich and abundant.

In recent years, Fengjing Town has taken many measures to advocate the integration of culture and tourism, creating a new experience of intangible cultural heritage, and has organized thematic activities such as "Shopping Festival of Intangible Cultural Heritage in Ancient Town" and "Ancient Town Cultural Street , Intangible Cultural Heritage Bazaar", which have given the ancient town a deeper cultural connotation. Walking in

Fengjing, you can stroll along the corridor bridge, cook wine and sip tea to relive the leisurely life in the Jiangnan water town.

Immovable Cultural Relics Resources

In September 2005, Fengjing Town was announced as one of the second batch of national famous historical and cultural towns, in which the area of historical and cultural landscape protection zone is 104.1 hectares and the area of core protection zone is 31.24 hectares. There are 44 immovable cultural relics in the town, including 3 Shanghai city-level protected sites, 7 district-level protected sites and 34 district-level protected places.

枫泾消防纪念塔及
东区火政会

Fengjing Fire Fighting
Memorial Tower and East
District Fire Station

年代：近现代
类别：近现代重要史迹及代表性建筑
保护级别：上海市文物保护单位
利用情况：开放参观
Era: Modern times
Category: Modern important historical sites and
representative buildings
Conservation level: Shanghai city-level protected site
Utilization: Open for visit

枫泾消防纪念塔建于 1926 年，是松善枫泾救火联合会为纪念义务救火亡故人员而建立的纪念塔。塔高 7 米，上书"松善枫泾救火联合会历届故亡同志纪念塔"，塔首呈鼓形，上饰四环，下为救火联合会徽标，是上海郊区至今唯一保存完好的消防历史纪念塔。东区火政会建于 1937 年，是上海地区仅存的较为完整的近代消防旧址。坐西南朝东北，砖木结构，硬山顶，穿斗式结构，六椽架，总占地面积 185 平方米，建筑面积 114 平方米。

Built in 1926, Fengjing Fire Fighting Memorial Tower was erected by Songshan Fengjing Fire Fighting Association to commemorate the deaths of volunteer firefighters. The tower is 7 meters high, with the inscription "Memorial Tower for the deceased comrades of Songshan Fengjing Fire Fighting Association". The head of the tower is in the shape of a drum, decorated with four rings on the top and the logo of the Fire Fighting Association underneath. It is the only well-preserved memorial tower of fire-fighting history in the suburbs of Shanghai so far. Eastern Fire Administration Council was built in 1937, which is the only surviving relatively complete site showing the former modern firefighting system in the Shanghai area. The whole building is a brick and wood structure facing the northeast. Using a flush gable roof (a term used in the construction industry to describe a hard covering material used as a roof), it is a column and tie construction (a structural system commonly used in architectural and bridge designs. It primarily utilizes columns and connected horizontal members ties to bear the structural loads and transfer them to the foundation or supporting structures) with six rafters, covering a total area of 185 square meters and a floor area of 114 square meters.

人民公社旧址
Former Site of the People's Commune

年代: 近现代
类别: 近现代重要史迹及代表性建筑
保护级别: 上海市文物保护单位
利用情况: 开放参观
Era: Modern times
Category: Modern important historical sites and
representative buildings
Conservation level: Shanghai city-level protected site
Utilization: Open for visit

　　人民公社旧址建于 1959 年 7 月，坐北朝南，砖木结构，共 7 间，总占地面积 2208 平方米，建筑面积 219 平方米。1959 年至 1984 年，"枫围人民公社"在此办公 25 年。1984 年起，枫围乡人民政府在此办公，直至 1993 年与枫泾镇 "乡镇合一"，后旧址被用作文化站，现为开放景点。

　　The former site of the People's Commune was built in July 1959. The building faces south, and it is a brick and wood structure with 7 rooms, covering a total area of 2208 square meters and a floor area of 219 square meters. From 1959 to 1984, the "Fengwei People's Commune" worked here for 25 years. Since 1984, the People's Government of Fengwei Township had been working here until 1993 when it was united with Fengjing Township as one. Afterwards, the former site was used as a cultural station, which is now an open attraction.

朱学范故居

Zhu Xuefan Former Residence

年代：清代
类别：近现代重要史迹及代表性建筑
保护级别：上海市文物保护单位
利用情况：开放参观
Era: Qing Dynasty
Category: Modern important historical sites and representative buildings
Conservation level: Shanghai city-level protected site
Utilization: Open for visit

朱学范，杰出爱国民主战士和政治活动家。其故居建于清光绪三十一年（1905）以前，坐西南朝东北，主要由二层砖木结构楼房及单间平房等组成，主体布局呈"凸"字形，占地面积632平方米，建筑面积582平方米。前幢硬山顶，面宽5间，进深八椽架，明次间抬梁，落地格栅门、格栅窗。故居内陈列有朱学范生平展。

Zhu Xuefan was an outstanding patriotic democratic fighter and political activist. His former residence was built before the 31st year of the reign of Guangxu in the Qing Dynasty (1905). Facing northeast, it mainly consists of two-story brick and wood structure building and one-room bungalow, etc. The main layout is in the shape of " 凸 ", covering an area of 632 square meters and a floor area of 582 square meters. The front building has flush gabled roof, 5 bays, eight-rafter-length depth. It has strut-beam in central bay and secondary bays, floor-to-ceiling partition doors and grille windows. Zhu Xuefan life exhibition is displayed therein.

程十发祖居
Cheng Shifa Ancestral Residence

年代: 清代
类别: 近现代重要史迹及代表性建筑
保护级别: 金山区文物保护单位
利用情况: 开放参观
Era: Qing Dynasty
Category: Modern important historical sites and representative buildings
Conservation level: Jinshan district-level protected site
Utilization: Open for visit

　　程十发,中国海派书画画家。其祖居建于清代,坐东北朝西南,砖木结构,前幢为平房,后幢为两层楼房,共5间,占地面积446平方米。祖居内恢复了程十发祖父、父亲行医的诊所厅堂和程十发出生居住的卧室。卧室里雕花床、梳妆台一应俱全。同时,祖居内还展出了程十发部分画作以及生活、创作用具,为研究程十发早期在枫泾的活动提供了重要的实证载体。

Cheng Shifa was a Chinese painter and calligrapher of the Shanghai School. His ancestral residence was built with a brick and wood structure in the Qing Dynasty, and there are 5 rooms in the two buildings. The front building is a bungalow and the back building is a two-story building, covering an area of 446 square meters. Inside the house, the clinic hall where Cheng Shifa's grandfather and father used to practice medicine and the bedroom where Cheng Shifa was born have been restored. The bedroom is fully equipped with carved beds and dressing tables. In addition, some of Cheng Shifa's paintings as well as his living and writing utensils are also on display, providing important evidence for the study of Cheng Shifa's early activities in Fengjing.

蔡以台宅

Cai Yitai Residence

年代：清代
类别：古建筑
保护级别：金山区文物保护点
利用情况：其他用途
Era: Qing Dynasty
Category: Ancient architecture
Conservation level : Jinshan district-level protected place
Utilization: Other uses

蔡以台，清代状元，善辨金石、图书等文物真伪，存有《三友斋样遗稿》《姓氏窃略》。蔡以台宅始建年代不详，属清代风格，坐西北朝东南，砖木结构，三进院落，共占地约 852 平方米，建筑面积 500 平方米。第一、三幢为二层楼房，第二幢为平房，硬山灰瓦顶，穿抬混合式结构。三进东北侧有封火墙隔断。

Cai Yitai, one of the top scorers in the palace examination of the Qing Dynasty, is good at distinguishing the authenticity of cultural relics such as bronze and stone inscriptions, books, etc. He has six volumes of *Sanyouzhai Sample Postscripts* and *Xingshi Qie Lüe* (a book that lists and explains various Chinese surnames). Cai Yitai Residence was built in an unknown era, but belongs to the style of the Qing Dynasty. Facing southeast, it is a brick and wood structure of three courtyards, covering a total area of about 852 square meters, with a floor area of 500 square meters. The first and third buildings are two-story houses. The second one is a bungalow with flush gable roof covered with gray tiles and is a mixed column-and-tie and post-and-lintel (a structural system used in architecture and construction that involves vertical posts supporting horizontal lintels or beams. This system has been used since ancient times and is one of the oldest methods of constructing buildings) structure. The northeast side of the third one is separated by a firewall (fire protection measure).

郁家祠堂
Yu Family Ancestral Hall

年代: 清代
类别: 古建筑
保护级别: 金山区文物保护点
利用情况: 商业用途
Era: Qing Dynasty
Category: Ancient architecture
Protection level: Jinshan district-level protected place
Utilization: Commercial use

　　郁家祠堂建于清雍正四年（1726），坐东北朝西南，砖木结构。由前厅、后厅及左右二厢房组成，中间为天井，四合院式布局，占地面积 821 平方米，建筑面积 520 平方米。前后厅为抬梁式硬山灰瓦顶。前厅山墙两侧有观音兜，二厢房及后厅檐枋上雕有"福禄寿"及花卉等图案。

　　The Yu Family Ancestral Hall was built in the 4th year of the Yongzheng reign of the Qing Dynasty (1726). Facing the southwest, the brick and wood structure consists of a front hall, a rear hall and two side rooms. With a patio in the middle, the layout follows the traditional courtyard style, covering an area of 821 square meters and a building area of 520 square meters. The front and rear halls have a column and strut-beam construction with a flush gable roof covered with gray tiles. On the wall of the front hall, there are Guanyin gable wall (a gable wall with a curved profile projecting beyond the wall) on both sides. The eave tiebeams of the two side rooms and the rear hall are carved with patterns such as "Fu Lu Shou" (Good Fortune, Prosperity, Longevity) and flowers.

致和桥
Zhihe Bridge

年代: 元代
类别: 古建筑
保护级别: 金山区文物保护单位
利用情况: 其他用途
Era: Yuan Dynasty
Category: Ancient architecture
Conservation level: Jinshan district-level protected site
Utilization: Other uses

致和桥俗称圣堂桥，元致和元年（1328）建，东西走向，跨枫泾市河。桥全长 26.3 米，宽 2.9 米，拱径 7.5 米，是金山区保存较为完整的元代石拱桥。

Zhihe Bridge, commonly known as Shengtang Bridge, was built in the 1st year of Zhihe reign of the Yuan Dynasty (1328). It spans the Fengjing Shihe River in an east-west direction. The bridge measures 26.3 meters in length, 2.9 meters in width, and the arch has a diameter of 7.5 meters. It is a well-preserved stone arch bridge of the Yuan Dynasty in Jinshan District.

宝源桥
Baoyuan Bridge

年代: 清代
类别: 古建筑
保护级别: 金山区文物保护单位
利用情况: 其他用途
Era: Qing Dynasty
Category: Ancient architecture
Conservation level: Jinshan district-level protected site
Utilization: Other uses

宝源桥，又名义丰桥，建于清代，花岗石质，立壁柱式三跨梁桥。西南至东北走向，跨枫泾下西市河，全长 15.6 米，宽 2.31 米，净跨 10.6 米。桥面由 12 块长条石拼建，桥身两侧刻有正楷阳文"重建宝源桥"，护栏板由砖石砌成。

Baoyuan Bridge, commonly known as Yifeng Bridge, was built during the Qing Dynasty. It is a three-span beam bridge made of granite, with pillar-style abutments, spanning the Fengjing Xiaxishi River in a southwest to northeast direction. The bridge measures 15.6 meters in length, 2.31 meters in width, and the net span is 10.6 meters. The bridge deck is constructed with 12 long stone slabs, and both sides of the bridge are engraved with the words "重建宝源桥"(Reconstruction of Baoyuan Bridge) in regular embossed script. The guardrail plate is made of masonry.

朱家角镇

1 放生桥
朱家角镇北大街东首

2 课植园
朱家角镇西井街 111 号

3 朱家角城隍庙
朱家角镇漕河街 69 号

4 大清邮局旧址
朱家角镇西湖街 35 号

5 东井茶楼
朱家角镇东井街 122 号

6 童天和国药号
朱家角镇大新街 60 号

7 涵大隆酱园
朱家角镇北大街 287 号

8 席家住宅
朱家角镇东湖街 49 弄 42 号

Zhujiajiao Town

1 *Fangsheng Bridge*
East end of North Street, Zhujiajiao Town

2 *Kezhi Garden*
No.111, Xijing Street, Zhujiajiao Town

3 *Zhujiajiao Chenghuang Temple*
No.69, Caohe Street, Zhujiajiao Town

4 *Former Site of the Qing Dynasty Post Office*
No.35, Xihu Street, Zhujiajiao Town

5 *Dongjing Tea House*
No.122, Dongjing Street, Zhujiajiao Town

6 *Tongtianhe Chinese Pharmacy*
No.60, Daxin Street, Zhujiajiao Town

7 *Handalong Sauce and Pickle Shop*
No.287, North Street, Zhujiajiao Town

8 *Xi Family Residence*
No.42, Lane 49, Donghu Street, Zhujiajiao Town

图例 LEGENDS

全国重点文物保护单位
National priority protected site

上海市文物保护单位
Shanghai city-level protected site

区级文物保护单位
District-level protected site

区文物保护点
District-level protected place

其他景点
Other attractions

历史风貌区范围
Scope of historical district

游览路线
Touring route

大淀湖 Dadian Lake

课植园
Kezhi Garden

放生桥
Fangsheng Bridge

东井茶楼
Dongjing Tea House

圆津禅院
Yuanjin Zen Temple

泰安桥
Taian Bridge

涵大隆酱园
Handalong Sauce and Pickle Shop

朱家角城隍庙
Zhujiajiao Chenghuang Temple

童天和国药号
Tongtianhe Chinese Pharmacy

大清邮局旧址
Former Site of the Qing Dynasty Post Office

席家住宅
Xi Family Residence

淀浦河 Dianpu River

西湖街 Xihu Street

东湖街 Dongshu Street

东市街 Dongshi Street

人和路 Renhe Road

祥凝浜路 Xiangningbang Road

朱泖河 Zhumao River

珠溪路 Zhuxi Road

沪青平公路 Huqingping Highway

漕家浜 Caojia\Shihe River

本图为位置示意，与实际尺寸不符
Illustration is not proportional to the actual scale

朱家角镇
Zhujiajiao Town

交通指南:
地铁 17 号线至朱家角站。

Transportation Guide:
Metro Line 17 to Zhujiajiao Station.

江南明珠，烟火繁华

朱家角镇位于上海西侧的淀山湖之畔，素有"江南明珠"之称。镇上河多、桥多、明清建筑多、文化遗迹多，街市繁荣，沿街两侧的大小商号鳞次栉比，全盛时期有千家之余，至今仍有不少百年老店。走进老弄深巷，曲径通幽，大户深藏，名人贤士辈出，人文底蕴深厚。

朱家角大约成陆于7000年前，6000多年前即有先民在此繁衍生息。三国时期，这里已有村落，宋元期间形成集市，名朱家村。后因水运便利，商业日盛，朱家角先是成为棉花、棉布的交易市场，后又成为大米和食油的集散地，是方圆数十里的经济中心。如《珠里小志》宋如林撰序中所言："今珠里为青溪一隅，烟火千家……其街衢绵亘，商贩交通，水木清华，文儒辈出……过是里者，群羡让耕让畔之风犹古，而比户弦歌之不辍也"，朱家角镇因水聚民，依水成市，是江南古镇中的一颗璀璨明珠。

朱家角镇的漕港河、泖河及附近湖泊与河水交织形成五里、七里一纵浦，七里、十里一横塘的便捷河网系统。镇上有9条临水而建的长街，诸多明清建筑傍水而建，古风犹存。北大街东起放生桥、西至美周弄，被称为"上海市郊保存得最完整的明清建筑第一街"。数十座古桥犹如镶嵌在碧波长卷上的明珠，古朴典雅。

朱家角镇的建筑大多是灰瓦白墙，只有少数民国以来曾修缮过的房屋盖上了红色"洋瓦"。集商业、居住、生产于一体的民居建筑设计很有特点。

水巷两岸的房屋多建有副檐，为底层房屋遮雨、防晒，吊空的副檐还可供河边停船遮风避雨。两岸的房屋不但紧靠，还建有许多气楼，俗称"老虎窗"，可以提高房屋的使用价值。水巷两岸有着千姿百态的河埠、缆船石、灯龛，设计精巧实用。

朱家角镇有国家级非遗项目"吴歌"（青浦田山歌），有"摇快船""涵大隆酱园酱菜制作技艺""船拳""淀山湖传说"4个上海市级非遗项目，有"牛角镗舞""童天和药号膏方文化""稻草编结制作技艺"3个青浦区级非遗项目。

如今的朱家角镇已成为上海乃至长三角地区最具人气的旅游地之一，来自天南地北的游客为古镇带来了无尽的生机与繁荣。漫步古镇，文艺场馆遍布大街小巷，古建筑可以成为新艺术的演出地，现代建筑也可以成为古老文化的展示厅，处处洋溢着浓郁的艺术气息。

不可移动文物资源

2007 年 5 月，朱家角被公布为第三批中国历史文化名镇，历史文化风貌保护区面积为 179.67 公顷，核心保护区面积为 34 公顷。镇域内现有各级不可移动文物 84 处，其中上海市文物保护单位 2 处，区级文物保护单位 12 处，区文物保护点 70 处。

A town as the Pearl of Jiangnan, bustling with the hustle and bustle

Zhujiajiao Town is on the shore of Dianshan Lake on the west side of Shanghai, known as the "Pearl of Jiangnan". The town is rich in rivers, bridges, buildings of the Ming and Qing dynasties and cultural relics. As a prosperous market, large and small shops line up along both sides of the street, where there were more than a thousand in its heyday and still are quite a number of century-old stores. Walking into the old lanes and deep alleys, the winding paths will lead to hidden large households where the birthplace of many celebrities and scholars are. The town has a deep humanistic heritage.

Zhujiajiao is believed to have become land around 7000 years ago, with the earliest inhabitants dating back over 6000 years. During the Three Kingdoms period, there were already villages here, and during the Song and Yuan dynasties, a market formed to be known as Zhujiacun. Due to its convenient water transportation, commerce thrived in Zhujiajiao. It became the economic center for dozens of li (1 li is approximately 500 meters) around, first as a trading market for cotton and cotton cloth, and later as a distribution center for rice and cooking oil. As described in the preface to "Zhuli Records" by Song Rulin, Today, Zhuli is but a corner of Qingxi, with bustling streets and countless households...Its streets are long and

---

朱家角镇参观指南 Zhujiajiao Town Visiting Guide

游览路线：
课植园→东井茶楼（阿婆茶楼）→放生桥→泰安桥→圆津禅院→涵大隆酱园→朱家角城隍庙→童天和国药号→席家住宅→大清邮局旧址

古镇美食特产：
熏青豆、酱油虾、烧卖、粽子、扎蹄、扎肉、五香豆腐

Tourist Route：
Kezhi Garden → Dongjing Tea House (Apo Tea House) → Fangsheng Bridge → Taian Bridge → Yuanjin Zen Temple → Handalong Sauce and Pickle Shop → Zhujiajiao Chenghuang Temple → Tongtianhe Chinese Pharmacy → Xi Family Residence → Former Site of the Qing Dynasty Post Office

Local Specialties：
Smoked Green Beans, Soy Sauce Shrimp, Shaomai Dumplings, Zongzi (sticky rice dumplings), Braised Pig Hooves, Braised Pork, Five-Spice Tofu

winding, bustling with merchants, surrounded by clear waters and lush woods, producing numerous scholars and literati... Passing through this place, one cannot help but admire its ancient traditions that continue to thrive, like the way of alternating the use of fields, and the unceasing melodies of singing and playing instruments. Having gathered people with the water and became the town along the water, Zhujiajiao is a bright pearl in the ancient town of Jiangnan.

The Caogang River, Maohe River and nearby lakes and rivers in Zhujiajiao Town are intertwined to form a convenient water network system with a longitudinal river of five or seven li, and a transverse pond of seven or ten li. There are 9 long streets built along the water, with many Ming and Qing dynasty buildings still standing, preserving the ancient atmosphere. The North Street stretches from Fangsheng Bridge in the east to Meizhou Lane in the west, and is known as the "first street with the most intact preservation of Ming and Qing dynasty buildings in the suburbs of Shanghai". Dozens of ancient bridges are like pearls embedded in the meandering blue waves, exuding a sense of simplicity and elegance.

The buildings in Zhujiajiao are mostly characterized by gray tiles and white walls, with only a few houses that have been renovated since the Republic of China Period using red "Western tiles". The architectural design of the residential buildings, which combine commerce, residence, and production, is quite distinctive. Many houses along the water lanes have secondary

roofs to shelter the lower-level houses from rain and sun, and the suspended eaves can also provide shelter for boats along the river. The houses on both sides of the water lanes are not only closely adjacent to each other, but also have many "Qi Lou" (Qi Lou is a common architectural feature in ancient buildings, typically located at a high position. It usually consists of multiple small windows that can be opened or closed to adjust airflow), commonly known as "tiger windows", which can increase the utilization value of the houses. The water lanes on both sides also have various riverfront docks, rock-cable boats, and lantern niches, all of which are designed with both beauty and practicality.

Zhujiajiao has a national-level intangible cultural heritage project called Wu Ge (Farm Song of Qingpu). It also has four Shanghai-level intangible cultural heritage projects, including "Yao Kuai Chuan" (a boat-rocking activity), "Handalong Sauce and Pickle Shop Fermented Pickle Making Art", "Chuan Quan" (a form of martial arts), and "Dianshan Lake Legend". Additionally, there are three intangible cultural heritage projects at the Qingpu district-level, which are "Niu Jiao Tang Dance" (bull-horn dance), medicinal plaster of Tongtianhe Chinese Pharmacy, and "Dao Cao Bian Jie Making Art" (straw weaving art).

Nowadays, Zhujiajiao Town has become one of the most popular tourist destinations in Shanghai and even in the Yangtze River Delta region. Visitors from all over the country bring endless vitality and prosperity to this ancient town. Taking a leisurely stroll through the town, you will find art venues scattered along the streets and alleys. The ancient buildings have become stages for new forms of art, while modern architecture serves as exhibition halls for ancient culture. Everywhere you go, there is a strong artistic atmosphere permeating the air.

Immovable Cultural Relics Resources

In May 2007, Zhujiajiao Town was announced as the third batch of national famous historical and cultural towns. The area for the protection of historical and cultural features covers 179.67 hectares, with a core protection area of 34 hectares. Currently, there are 84 immovable cultural relics at various levels within the town, including 2 Shanghai city-level protected sites, 12 district-level protected sites and 70 district-level protected places.

放生桥
Fangsheng Bridge

年代: 清代
类别: 古建筑
保护级别: 上海市文物保护单位
利用情况: 开放参观
Era: Qing Dynasty
Category: Ancient architecture
Conservation level: Shanghai city-level protected site
Utilization: Open for visit

放生桥始建于明朝隆庆五年（1571），由慈门寺僧人性潮募款筹建，清嘉庆十七年（1812）重建，是上海地区现存最大、最长、最高的五孔石拱桥。

放生桥横跨于朱家角镇东首的漕港河（淀浦河）上，造型精巧、气势宏伟，全长72米，宽5米，高8.65米，中孔径距13米，二孔径距7.5米，三孔径距5.4米，素有"井带长虹"之称。

放生桥的桥身雄伟稳固，弧度优美，桥上石刻技艺高超。桥南堍东侧建有碑亭，存碑三通。临水筑以石驳，凿有锁缆孔，为舟楫停泊所用。

The Fangsheng Bridge was initially built in the 5th year of the Longqing reign of the Ming Dynasty (1571) through fundraising efforts by the monk Xingchao of Cimen Temple, and later rebuilt in the 17th year of the Jiaqing reign of the Qing Dynasty (1812). It is the largest, longest, and highest five-arch stone bridge in the Shanghai area that remains in existence today.

The Fangsheng Bridge, which spans the Caogang River (Dianpu River) at the eastern end of Zhujiajiao Town, is delicately designed and grand in scale. It has a total length of 72 meters, a width of 5 meters, and a height of 8.65 meters. The diameter of the central arches is 13 meters, while the diameter of the secondary arches is 7.5 meters, and the diameter of the thirdary arches is 5.4 meters. It is known as the bridge of "the well with a jade belt-length and rainbow-shaped".

The bridge's structure is majestic and sturdy, with a gracefully curved arch. The stone carvings on the bridge exhibit exceptional craftsmanship. On the east side of the southern end of the bridge, there is a stele pavilion with three inscribed stele. Stone revetments are built by the water, with holes carved for mooring ropes, serving as a docking area for boats.

课植园
Kezhi Garden

年代: 近现代
类别: 近现代重要史迹及代表性建筑
保护级别: 青浦区文物保护单位
利用情况: 开放参观
Era: Modern times
Category: Modern important historical sites and representative buildings
Conservation level: Qingpu district-level protected site
Utilization: Open for visit

　　课植园是古镇上最大的庄园式园林建筑，由厅堂区、假山区、园林区三部分构成，有门厅、客厅、堂楼、迎贵厅、耕九余三堂、藏书楼、望月楼、打唱台、倒挂狮子亭、碑廊、双眼井、九曲桥、课植桥等各种建筑及生活用房 200 余间，布局错落有致、独具匠心。

　　The Kezhi Garden is the largest manor-style garden architecture in the ancient town, consisting of three parts: the hall area, the rockery area, and the garden area. It includes more than 200 various buildings and living quarters, such as gate hall, reception hall, main building, welcome hall, Gengjiuyusan Hall, book collection house, Wangyue Tower, stage for performing Chinese opera, Daoguashizi Pavilion, stele corridor, double-hole well, zigzag bridge, and Kezhi Bridge. The layout is well-arranged and unique.

朱家角城隍庙

Zhujiajiao Chenghuang Temple

年代：清代
类别：古建筑
保护级别：青浦区文物保护单位
利用情况：宗教活动
Era: Qing Dynasty
Category: Ancient architecture
Conservation level: Qingpu district-level protected site
Utilization: Religious activities

朱家角城隍庙原在镇南的雪葭浜，清乾隆二十八年（1763），徽州人程履吉迁至今址。庙宇布局雅致，原有十二胜景，中间为头门、戏台、大殿，两侧为庑，左边有寅清堂、熙春台、假山方池、梅亭、玉照廊、月香室，右边有凝和书屋、荷净山房、潭影阁、荷花池、可娱斋、乐溪庐、挹秀轩、花神殿、怡亭、含清榭，旁边有小曲溪，上架小石桥，桃李杨柳随处可见。1964 年 1 月，后殿、寝宫焚毁。至 1985 年，庙中亭台楼阁、假山水池已无存，头门、戏台、两庑、大殿等主体建筑保存基本完好。

The Chenghuang Temple (a traditional Chinese temple dedicated to the City God, who is believed to protect the local area and its people) in Zhujiajiao was originally located in Xuejia creek at the south of the town. In the 28th year of the Qianlong reign of the Qing Dynasty (1763), a person named Cheng Lvji from Huizhou relocated it to its current location. The layout of the temple is elegant, with 12 main attractions. In the center, there is the main gate, a stage, and a main hall. On the sides, there are wing rooms. On the left side, there are Yinqing Hall, Xichun Terrace, a rockery square pond, Meiting Pavilion, Yuzhao Corridor, Yuexiang Room, and on the right side, there are Ninghe Study, Hejing Shanfang (Lotus Purification Mountain House), Tanying Pavilion, Lotus Pond, Keyu Studio, Lexi Hut, Yixiu Pavilion, Huashen Hall, Yi Pavilion, and Hanqing Xie (House built on a platform). Next to it, there is a small creek under small stone bridge, and peach, plum, willow trees can be seen everywhere. In January 1964, the rear hall and the bedchambers were burned down. By 1985, the pavilions, towers, rockeries, and water ponds in the temple had disappeared, but the main buildings such as the main gate, stage, wing rooms, and main hall remained relatively intact.

大清邮局旧址

Former Site of the Qing Dynasty Post Office

年代: 清代
类别: 近现代重要史迹及代表性建筑
保护级别: 青浦区文物保护单位
利用情况: 开放参观
Era: Qing Dynasty
Category: Modern important historical sites and representative buildings
Conservation level: Qingpu district-level protected site
Utilization: Open for visit

　　朱家角大清邮局始建于清同治年间（1862—1874），初为民间信局。光绪二十二年（1896），清廷开办国家邮政，遂成为大清邮政官局邮寄代办所，是上海地区清代 13 家通邮驿站之一。

　　建筑坐南朝北，二层砖木结构，三开间，外立面为仿欧洲式样，占地面积约 70 平方米。门面中间是两扇大门，门上方镶有"朱家角邮局"门额，门额上下都有图案花饰，两侧各有一扇拱形窗户，大门外侧有一个仿清代铜制铸龙邮筒。

　　The former site of the Qing Dynasty Post Office in Zhujiajiao was initially established as a private communication station during the Tongzhi reign of the Qing Dynasty (1862–1874). In the 22nd year of the Guangxu reign (1896), the imperial court established a national postal service, and it became an official postal agency of the Qing Dynasty in the Shanghai area, handling mail services. It was one of the 13 regular postal stations during the Qing Dynasty.

　　The building faces north and is a two-story brick and wood structure. It has three bays and features a European-style exterior façade, covering an area of approximately 70 square meters. In the center of the storefront are two large doors, above which is an inscription that reads "Zhujiajiao Post Office", adorned with decorative patterns. On each side of the doors, there is an arched window. Outside the main entrance, there is a bronze-cast dragon-shaped mailbox, imitating the style of Qing Dynasty postal mailboxes.

东井茶楼
Dongjing Tea House

年代: 清代
类别: 古建筑
保护级别: 青浦区文物保护点
利用情况: 开放参观
Era: Qing Dynasty
Category: Ancient architecture
Conservation level: Qingpu district-level protected place
Utilization: Open for visit

　　东井茶楼建于清末，曾做过大公义米行商业店铺，现为茶楼。坐南朝北，砖木结构，前后两埭，东西两面相连。南楼三开间，面阔 10.5 米，进深 8 米，七架梁，歇山顶，上铺小青瓦；南面上下两层每间均有 6 扇花结嵌玻璃槛窗，北面有砖雕仪门。北楼面阔 10.5 米，进深 5.1 米，五架梁，硬山顶，两面有观音兜封火墙，前后有廊，廊下有宫式挂落。

Dongjing Tea House, built in the late Qing Dynasty, was once a commercial store called Dagong Fair Rice Store, but now a tea house. The building faces north, and has a brick and wood structure. The building consists of front buildings and back buildings, which is connect on both sides of the east and west. The south building has a span of three bays, 10.5 meters wide and 8 meters deep, supported by 7-purlin beams, and a gable and hip roof covered with small gray tiles. Both the upper and lower levels of the south side have six windows with flower-embedded glass in each room, while the north side has a brick-carved ceremonial gate. The north building is 10.5 meters wide and 5.1 meters deep, with 5-purlin beams and a flush gable roof. It has Guanyin gable wall on both sides and colonnades in the front and back, with palace-style hanging ornaments in the galleries.

童天和国药号

Tongtianhe Chinese Pharmacy

年代: 清代
类别: 古建筑
保护级别: 青浦区文物保护单位
利用情况: 商业用途
Era: Qing Dynasty
Category: Ancient architecture
Conservation level: Qingpu district-level protected site
Utilization: Commercial use

童天和国药号是朱家角历史最悠久的药号，由清代宁波富绅童氏独资创设。店堂坐东朝西，二层二进不规则结构，房屋向东减小，前、后埭间各有小天井一处，占地面积200余平方米。

临街店门面是高高的封火墙，中间置石库门，石库门上方有"童天和药号"五个金字。前埭二开，七架梁；二埭，七架梁；三埭为三层结构，房屋内木结构雕刻精美。

药号内保存着旧式的柜台、药具、药匮，具有传统的国药号风格风貌。"童天和"药材道地、加工考究，尤以精工炮制的饮片享负盛名。

朱家角多位著名国医，如金乃声、叶志廉等都曾在此坐堂处方。名医配良药，使"童天和"声名远扬。

Tongtianhe Chinese Pharmacy is the oldest pharmacy in Zhujiajiao, established by the wealthy Ningbo merchant Tong family during the Qing Dynasty. The building faces west and has an irregular two-story, two-yard layout. It gradually decreases in size towards the east, with small courtyards at the front and back, covering an area of over 200 square meters.

The storefront facing the street features a tall firewall, with a Shikumen in the middle, above which are five golden characters that read "Tongtianhe Chinese Pharmacy". The first building has two bays with 7-purlin beams, the second building has 7-purlin beams (a beam that supports seven purlins), and the third building is a three-story building with exquisitely carved wooden decorations inside.

The pharmacy preserves traditional counters, medicinal utensils and medicine cabinets, embodying the traditional style and appearance of a traditional Chinese pharmacy. Tongtianhe is known for its authentic medicinal herbs and meticulous processing, especially its renowned herbal slices.

Many famous traditional Chinese doctors in Zhujiajiao, such as Jin Naisheng and Ye Zhilian, have practiced and prescribed medicine at this pharmacy. The combination of skilled doctors and high-quality medicines has earned Tongtianhe a reputation for excellence.

涵大隆酱园

Handalong Sauce and Pickle Shop

年代：近现代
类别：近现代重要史迹及代表性建筑
保护级别：青浦区文物保护单位
利用情况：商业用途
Era: Modern times
Category: Modern important historical sites and representative buildings
Conservation level: Qingpu district-level protected site
Utilization: Commercial use

涵大隆酱园创建于清末民初，为典型的前店后作坊建筑。酱园坐东朝西，砖木结构，现存有前埭、厢房、后埭，占地面积 250 平米。其门墙为高大的封火墙，入口为石库门，设黑漆铁钉双开大门，砖雕匾状招牌位于店门正上方，墙上"酱园"二字白底黛字。进门过天井，是前埭屋，二层，底层落地长窗，地面铺设方砖。一层为店堂，木质柜台和货架上放满了各种瓶装和散装的酱菜，二层为卧室。后埭屋二层，底层是酿造工厂，二层是仓库。厢房面宽三开间，二层建筑，其挑廊与前后埭屋的挑廊连成回廊，廊间为第二天井。该建筑的梁下有精美雕饰，檐下及落地长窗上均有雕花，施金色。涵大隆酱园所酿酱油曾在 1915 年巴拿马万国博览会获金奖。目前前店依然作酱园店铺。

Handalong Sauce and Pickle Shop was established in the late Qing Dynasty and early Republic of China. It is a typical front shop and rear workshop architecture. The sauce and pickle shop faces the west, with brick and wood structure. It currently has a front rooms, wing rooms and back rooms, covering an area of 250 square meters. The entrance of the Sauce Garden is a large firewall with a Shikumen, which is adorned with double-opening black painted gate with iron nails. Above the entrance, there is a brick-carved signboard in the shape of a plaque, with the words "Sauce and Pickle Shop" written on a white background in black ink. Upon entering the gate and passing through the courtyard, there is a front yard house, two-story high with floor-to-ceiling windows on the ground floor and square bricks on the floor. The first floor is a shop, with wooden counters and shelves filled with various bottled and bulk pickled vegetables, and the second floor is a bedroom. The back rooms is two-story high, with the lower floor being the brewing factory and the upper floor being the warehouse. The wing rooms are two-story three-bayed rooms. Its colonnades is connected to the galleries of the front and back rooms, forming a corridor, and there is a second courtyard between the corridors. The beams of the building are exquisitely carved, and there are carved patterns on the eaves and floor-to-ceiling windows, decorated in golden color. The soy sauce brewed by Handalong Sauce and Pickle Shop once won a gold medal at the 1915 Panama Pacific International Exposition. Currently, the front shop is still operating as a sauce and pickle shop.

席家住宅
Xi Family Residence

年代：明代
类别：古建筑
保护级别：青浦区文物保护单位
利用情况：开放参观
Era: Ming Dynasty
Category: Ancient architecture
Conservation level: Qingpu district-level protected site
Utilization: Open for visit

席家住宅建于明嘉靖年间（1522—1566）。席氏之祖为唐末大将席温，明清两代逐渐成为江南地区精于经商的望族。明代后期，席氏后裔自洞庭东山迁居于角里，至20世纪三四十年代，已历十代，有家谱可查者达27房数百人之多，涌现出《申报》创始人席裕祺、报业巨头席裕福、清代数学家席淦以及其他商界巨子和名人。

宅第坐南朝北，现存前厅、砖雕仪门，以及前后天井，占地面积229平方米。前厅三开间，面阔10.65米，九架梁，进深10.51米，硬山顶。前天井四周围墙顶砖砌雕花纹饰，厅后有轩。仪门门柱用水磨方砖贴面，门与外墙雕刻梅、兰、菊、牡丹、石榴、荷花、仙鹤、鹿、喜鹊、祥云、如意等吉祥图案。雕刻采用透雕、高浮雕、浅浮雕等技法，艺术价值极高。

The Xi Family Residence was built during the Jiajing period of the Ming Dynasty (1522–1566). The Xi family, whose ancestors can be traced back to General Xi Wen of the late Tang Dynasty, gradually became a well-respected and influential clan known for their business prowess in the Jiangnan region of China during the Ming and Qing dynasties. In the late Ming Dynasty, the descendants relocated from East Dongting Hill to Jiaoli and by the 1930s, they had reached their tenth generation, with a genealogical record of 27 branches and hundreds of members. Notable figures from the family include Xi Yuqi, the founder of the *Shenbao* newspaper; Xi Yufu, a newspaper tycoon; Xi Gan, a mathematician in the Qing Dynasty; and other successful figures in business and academia.

The residence faces north, with a total area of 229 square meters. The surviving parts of the residence include the front hall, a brick-carved ceremonial gate, and front and back courtyards. The front hall has a span of three bays, with a width of 10.65 meters and a depth of 10.51 meters, 9-purlin beams and a flush gable roof. The walls surrounding the front courtyard are adorned with intricate brick-carved patterns, and there is a veranda behind the hall. The pillars of the ceremonial gate are faced with water-grinded square bricks and are intricately carved with auspicious patterns such as plum blossoms, orchids, chrysanthemums, peonies, pomegranates, lotus, cranes, deer, magpies, auspicious clouds, and ruyi symbols. The carvings are done using techniques such as through-carving, high relief, and shallow relief, showcasing their high artistic value.

新场镇

Xinchang Town

1 新场第一楼书场
新场镇新场大街 424 号

1 *Xinchang First Building Storytelling Venue*
No.424, Xinchang Street, Xinchang Town

2 新场信隆典当
新场镇新场大街 367-371 号

2 *Xinchang Xinlong Pawnshop*
No.367-371, Xinchang Street, Xinchang Town

3 南山寺
新场镇新环南路 269 号

3 *Nanshan Temple*
No.269, Xinhuan South Road, Xindang Town

4 新场崇修堂
新场镇新场大街 350 号

4 *Xinchang Chongxiu Hall*
No.350, Xinchang Street, Xinchang Town

5 新场张氏宅第
新场镇新场大街 271 号

5 *Zhang Family Residence in Xinchang*
No.271, Xinchang Street, Xinchang Town

6 正顺（官）酱园
新场镇新场大街 137 号

6 *Zhengshun (official) Sauce and Pickle Shop*
No.137, Xinchang Street, Xinchang Town

7 千秋桥
新场镇洪东街 4 号

7 *Qianqiu Bridge*
No.4, Hongdong Street, Xindang Town

8 新场镇石驳岸及马鞍水桥
新场镇镇区

8 *Stone Revetment and Saddle-shaped Water Bridge in Xinchang*
Xinchang Town

沪南公路 Hunan Highway

千秋桥
Qianqiu Bridge

洪福桥
Hongfu Bridge

青龙桥
Qinglong Bridge

新场第一楼书场
Xinchang First Building
Storytelling Venue

三世二品坊
Three Generations
and Second-Rank Archway

新场信隆典当
Xinchang Xinlong
Pawnshop

新场崇修堂
Xinchang Chongxiu Hall

新场镇石驳岸及马鞍水桥
Stone Revetment and Saddle-shaped Water Bridge in Xinchang

新场张氏宅第
Zhang Family Residence in Xinchang

郑氏新宅
Zheng Family New Residence

中华楼
Zhonghua House

正顺（官）酱园
Zhengshun (official) Sauce and Pickle Shop

南山寺
Nanshan Temple

沪南公路 Hunan Highway

沪南公路 Sunzhong Road

新艺路 Xinyi Road

石笋街 Shisun Street

南五灶港 Nanwuzao Port

新环西路 Xinhuan West Road

笋南路 Sunnan Road

新场大街 Xinchang Street

新场港 Xinchang Port

奚家港 Qijiagang Road

新环南路 Xinhuan South Road

大治河 Dazhi River

新场镇
Xinchang Town

交通指南：
地铁 16 号线至新场站，换乘公交 1068 路。

Transportation Guide:
Metro Line 16 to Xinchang Station, transfer to Bus No.1068.

因盐而兴，活态新生

新场镇位于浦东新区南部，接长江之气、揽东海长风，是一座有着1300年历史的古镇，也是浦东规模最大、历史遗产最丰富的历史文化名镇。

古时新场，因盐而生、因盐而兴。新场成陆于唐代，俗呼"石笋滩"，后称"石笋里"。唐末，新场一带开始晒盐，成为"里镇"，并征收赋税。后汉乾祐年间，出现官办盐场。南宋建炎二年（1128），官府在新场以北10公里处设置盐监，名为"下砂盐场"，新场是该盐场的南场。此时盐场规模颇大，商贾入驻、市面繁荣。元代初年，下砂盐场向东迁移约十里，"石笋里"成为新的盐场所在地，于是便有了"新场"之名。清道光年间，大陆逐渐向海延伸，新场镇离海距离已达三十里，盐产量大幅下降，盐业风光不再。新场百姓便在芦苇田改种棉花、水稻，产业逐渐转向以粮棉桑麻为主，兼有家庭手工业，织布、绣花等传统手艺保存至今。

新场古镇依水而居，因河设市，门前连街市，窗外闻橹声，整个小镇如同一幅水墨长卷，民间曾有"十三牌楼九环龙，小小新场赛苏州"的美誉。从高处俯瞰，大片民居一户挨着一户，封火墙高高耸起，马头墙、观音兜互比高低，走在街上便能领略到宅、户、桥、巷纵横相连的玲珑格局。新场镇现存的主要街巷最早形成于元代，以新场大街、洪东街、洪西街为代表。水系是新场镇最具特色的文化景观，河道、驳岸、沿水街巷形成新场百姓水街相依的生活空间。洪桥港、包桥港、后市河、东横港将古镇划分为"井"

字形格局。新场镇西侧的后市河，小桥、流水、小宅院等要素展现了江南水乡小镇的风貌。现存石驳岸千余米，依然保留明清时期的风貌。驳岸上的缆船石孔及下水道造型别致，多座马鞍水桥保存现状良好，保留着传统蕴意。

新场镇现有 10 个项目被列入各级非物质文化遗产名录，还有非遗资源 80 多项。锣鼓书、浦东派琵琶已成为国家级非物质文化遗产，江南丝竹、灶花、卖盐茶已成为市级非物质文化遗产。上海本帮"老八样"传统菜肴以及"新场青""蜜露桃""椒桃片""海棠糕"等特色美食，历来广受称道。新场的石雕工艺、古玩陶瓷、书画装裱、白铁修理、特色土产、布店、茶楼、书场等老字号店铺鳞次栉比，民间手工匠人制作的竹编和木刻制品也琳琅满目，处处可寻。新场的小巷古街曲径通幽，碧水石桥遥相呼应，处处弥漫着世情温馨，铺陈着老浦东鲜活的生活画卷。

新场镇参观指南 Xinchang Town Visiting Guide

游览路线：
千秋桥→青龙桥→洪福桥→新场第一楼书场→新场信隆典当（新场历史文化陈列馆）→三世二品坊→新场崇修堂→新场镇石驳岸和马鞍水桥（后市河）→新场张氏宅第→郑氏新宅（法治文化博览园）→中华楼（锣鼓书艺术馆）→南山寺

古镇美食特产：
上海本帮"老八样"传统菜肴、蜜露桃、椒桃片、海棠糕、昂刺鱼咸肉菜饭

Tourist Route:
Qianqiu Bridge → Qinglong Bridge → Hongfu Bridge → Xinchang First Building Storytelling Venue → Xinchang Xinlong Pawnshop (Xinchang Historical and Cultural Exhibition Hall) → Three Generations and Second-Rank Memorial Archway → Xinchang Chongxiu Hall → Stone Revetment and Saddle-shaped Water Bridge in Xinchang (Houshi River) → Zhang Family Residence in Xinchang → Zheng Family New Residence (Rule of Law Culture Expo Park) → Zhonghua House (Luogushu Art Museum) → Nanshan Temple

Local Specialties:
Shanghai's "Old Eight Delicacies" Traditional Dishes, Honeyed Peaches, Spicy Peach Slices, Crabapple-shaped Cake, Vegetable Rice with Salted Meat and Yellow-headed Catfish

不可移动文物资源

　　2008 年 10 月，新场镇被公布为第四批中国历史文化名镇，历史文化风貌保护区面积 146.9 公顷，核心保护区面积 47.8 公顷。新场镇现有不可移动文物 81 处，为浦东新区街镇中不可移动文物最多的镇。其中上海市文物保护单位 2 处，区级文物保护单位 6 处，区文物保护点 73 处。镇内有 15 万平方米成片古建筑、60 余座古仪门、2000 米老街、1500 米明清石驳岸、4 条 "井" 字形河道、20 余座石桥、10 余座马鞍水桥，呈现着古街、古牌坊、古宅、古仪门、古桥、古驳岸、古寺、古银杏的 "八古" 历史风韵。

A town of salt production, vibrant and full of vitality

　　Xinchang Town is located in the southern part of Pudong New Area, embracing the spirit of the Yangtze River and the powerful winds of the East Sea. It is an ancient town with a history of 1300 years, making it the largest and most culturally rich historical town in Pudong.

　　In ancient times, Xinchang emerged and thrived due to salt production. It became a land during the Tang Dynasty, commonly known as "Shisun Tan", and later known as "Shisunli". In the late Tang Dynasty, salt drying began in the Xinchang area, making it a "Li Town" (an administrative division) subject to taxation. During the Qianyou reign of the Later Han Dynasty, government-operated salt fields started to appear. In the 2nd year of the Southern Song Dynasty's Jianyan reign (1128), the government set up a salt supervision office about 10 kilometers north of Xinchang, named "Xiasha Salt Field", with Xinchang serving as the southern field. At this time, the salt field had a significant scale, attracting merchants and thriving markets. In the early years of the Yuan Dynasty, the Xiasha Salt Field moved about ten li eastward, and "Shisunli" became the new location of the salt field, hence the name "Xinchang". During the Daoguang reign of the Qing Dynasty, the mainland gradually expanded towards the sea, and Xinchang was already thirty li away from the sea. Salt production declined sharply, and the salt industry lost its former glory. The people of Xinchang then switched to growing cotton and rice in the reed fields, gradually shifting their industries to focus on agriculture with crops like grain, cotton, mulberry, and hemp, alongside traditional handicrafts such as weaving and embroidery, which are still preserved today.

　　The ancient town of Xinchang is situated by the water, and its market was developed along the river. Street markets are right in front of the doors and the sound of paddles can be heard outside the windows. The entire town resembles

a scroll of ink painting. It was once praised as "Thirteen archways and nine rings of dragons, Xinchang, the small town that rivals Suzhou". From a higher vantage point, rows of houses are tightly packed together, with tall firewalls standing high. The Horse Head walls and Guanyin gable walls compete in height and elegance. Walking along the streets, one can appreciate the exquisite layout connecting residences, bridges, and alleys. The main streets and lanes of Xinchang Town were established during the Yuan Dynasty, represented by Xinchang Main Street, Hongdong Street, and Hongxi Street. The water system is the most distinctive cultural landscape of Xinchang Town. The rivers, revetments, and along-water streets form a living space where the locals rely on water. Hongqiao Port, Baoqiao Port, Houshi River, and Dongheng Port divide the ancient town into a " 井 "(well) shaped pattern. The Houshi River on the west side of Xinchang Town exhibits the charm of a typical water town in the Jiangnan region, with small bridges, flowing water and small courtyards. The existing stone revetments stretch over one kilometer, still retaining the appearance of the Ming and Qing dynasties. The unique and elegant cable ship stone holes and drainage system can be found on the revetments. Several well-preserved saddle-shaped water bridges still carry the traditional meanings.

Xinchang Town currently has 10 projects listed in various levels of intangible cultural heritage directories, with over 80 intangible cultural heritage resources. The traditional arts of drumming and storytelling, as well as the Pudong-style pipa, have been recognized as national-level intangible cultural heritage. The Jiangnan Sizhu or Jiangnan silk and bamboo music (traditional Chinese music: a form of musical performance in the Jiangnan region of China that combines the playing of silk and bamboo instruments. Silk instruments mainly include instruments such as pipa, erhu, and guzheng, while bamboo instruments include flutes and xiao. Jiangnan silk and bamboo music is known for its delicate and graceful playing style and is commonly used in folk music performances, opera accompaniment, and festive occasions), the Zaohua (a traditional rural celebration in China to mark a bountiful harvest) art and the Selling Salt and Tea (a kind of traditional folk dance) have been recognized as municipal-level intangible cultural heritage. The traditional dishes of Shanghai's local cuisine, such as the Old Eight Delicacies and the unique delicacies of Xinchangqing (a characteristic vegetable in Shanghai, also known as Xinchang Qingcai), Milutao (Honeyed Peaches: a traditional pastry in Shanghai made from white sugar, honey, and peach meat), Jiaotaopian (Spicy Peach Slices: a traditional cold dish in Shanghai made from fresh peaches, carrots, and green peppers, sliced after seasoning) and Haitanggao (Crabapple-shaped Cake: a traditional dessert in Shanghai made from glutinous rice flour, mung bean flour, and brown sugar) have always been well-known and highly praised. The old and well-established shops in Xinchang are lined up, showcasing various crafts such as stone carving, antique ceramics, calligraphy and painting mounting, iron restoration, specialty local products, fabric stores, tea houses, and storytelling venues. There is an abundance of bamboo weaving and wood carving products made by skilled craftsmen, readily available in every corner. The narrow alleys and ancient streets of Xinchang are tranquil and picturesque, with blue waters and stone bridges complementing each other. Everywhere is a warm and cozy atmosphere, presenting a vivid portrait of the lively life in old Pudong.

Immovable Cultural Relics Resources

In October 2008, Xinchang Town was announced as the fourth batch of national famous historical and cultural towns, with a protected area covering 146.9 hectares and a core protected area of 47.8 hectares. Xinchang Town currently has 81 immovable cultural relics, which is the highest number among all towns in Pudong New Area. Among them, there are 2 Shanghai city-level protected sites, 6 district-level protected sites and 73 district-level protected places. The town has 150 000 square meters of contiguous ancient buildings, over 60 ancient ceremonial gates, 2000 meters of old streets, 1500 meters of Ming and Qing Dynasty stone embankment, 4 cross-shaped waterways, over 20 stone bridges, and more than 10 saddle-shaped water bridges, presenting the historical charm of Eight Ancient elements such as ancient streets, ancient archways, ancient houses, ancient ceremonial gates, ancient bridges, ancient embankments, ancient temples, and ancient ginkgo trees.

新场第一楼书场
Xinchang First Building Storytelling Venue

年代： 清代
类别： 近现代重要史迹及代表性建筑
保护级别： 上海市文物保护单位
利用情况： 商业用途
Era: Qing Dynasty
Category: Modern important historical sites and representative buildings
Conservation level: Shanghai city-level protected site
Utilization: Commercial use

新场第一楼书场始建于清同治末年，20 世纪 20 年代翻建为三层楼房，时为镇上高度第一，故俗称 "第一楼"，30 年代起开设书场。占地面积 266.28 平方米，建筑面积 648.39 平方米，砖木结构，四坡顶，二、三层连排玻璃窗，5 架梁，临街。2004 年整体修缮，现建筑一层保留书场的使用功能，二、三层成为装修考究的茶楼。

Xinchang First Building Storytelling Venue was initially built at the end of the Qing Dynasty during the late Tongzhi reign. In the 1920s, it was reconstructed into a three-story building, becoming the tallest in town, hence known as the "First Building" by the locals. In the 1930s, it began operating as a storytelling venue. The building covers an area of 266.28 square meters, with a construction area of 648.39 square meters. It is a brick and wood structure, with a four-sloped roof. The second and third floors have a row of glass windows, supported by 5-purlin beams. The front of the building faces the street. In 2004, it underwent comprehensive renovation. The first floor is still used as a storytelling venue while the second and third floors have been transformed into a tastefully decorated teahouse.

新场信隆典当
Xinchang Xinlong Pawnshop

年代: 清代
类别: 近现代重要史迹及代表性建筑
保护级别: 上海市文物保护单位
利用情况: 开放参观
Era: Qing Dynasty
Category: Modern important historical sites and representative buildings
Conservation level: Shanghai city-level protected site
Utilization: Open for visit

　　新场信隆典当占地面积 1345.4 平方米，建筑面积 696 平方米。现存房屋为三进，前一进为平房，后二进为楼房，坐西朝东，砖木结构，硬山灰瓦顶，荷叶山墙。2006 年经过修缮后，由上海市历史博物馆设计布展成为"新场历史文化陈列馆"，馆内设有沧海桑田、煮海熬波、名人荟萃、生态古镇 4 个展厅。

　　Xinchang Xinlong Pawnshop is located on an area of 1345.4 square meters, with a building area of 696 square meters. The existing building is divided into three sections, with the front section being a single-story structure and the back two sections being multi-story. It faces east and is a brick and wood structure, with flush gable roof covered with gray tiles and lotus leaf-shaped walls. After renovation in 2006, it was redesigned and exhibited by the Shanghai History Museum as the "Xinchang Historical and Cultural Exhibition Hall". The exhibition hall consists of four themed halls: "Seas Changing into Mulberry Fields" "Cooking Sea and Boiling Waves" "Gathering of Celebrities" and "Ecological Ancient Town".

南山寺
Nanshan Temple

年代: 清代
类别: 古建筑
保护级别: 浦东新区文物保护单位
利用情况: 宗教活动
Era: Qing Dynasty
Category: Ancient architecture
Conservation level: Pudong New Area district-level
protected site
Utilization: Religious activities

　　南山寺由僧照建于元大德十年（1306），初名"常寂庵"。清顺治初年，由僧人九如重建，易名为南山禅寺。2005 年修缮后，现有房屋 6 幢，包括大雄宝殿、圆通殿等，总占地面积 6076.34 平方米，建筑面积 886.22 平方米，砖木结构，是新场重要的宗教活动场所。

Nanshan Temple, initially known as "Changji Temple" was built by the monk Sengzhao in the 10th year of the Yuan Dynasty's Dade reign (1306). In the early years of the Qing Dynasty, it was rebuilt by the monk Jiuru and renamed Nanshan Temple. After renovation in 2005, it now consists of six buildings, including the Mahavira Hall, the Yuantong Hall, etc., covering a total area of 6076.34 square meters with a building area of 886.22 square meters, and it is a brick and wood structure. It is an important religious site in Xinchang for various religious activities.

新场崇修堂
Xinchang Chongxiu Hall

年代: 清代
类别: 近现代重要史迹及代表性建筑
保护级别: 浦东新区文物保护单位
利用情况: 开放参观

Era: Qing Dynasty
Category: Modern important historical sites and representative buildings
Conservation level: Pudong New Area district-level protected site
Utilization: Open for visit

新场崇修堂由康碧梅于清光绪三十三年（1907）建造，占地面积426.6平方米。前后四进，一、二进为门面房和大厅，面阔3间；三、四进为内宅，皆二层楼，面阔3间，进深7米和6米，两侧各2间二层楼厢房，有雕刻精美的腰檐、木质栏杆和雀替。硬山屋面，小青瓦覆顶。院墙高广，山墙有观音兜和云寰两种饰件，造型优美。2021年6月打造为浦东院士风采馆暨新场院士之家。

Xinchang Chongxiu Hall covers an area of 426.6 square meters and was built by Kang Bimei in the 33rd year of the Guangxu reign of the Qing Dynasty (1907). It consists of four sections: the first and second sections are the storefront and the main hall, covering a width of 3 bays; The third and fourth sections are the living quarters, both two-story buildings covering a width of 3 bays, with a depth of 7 meters and 6 meters respectively. On both sides, there are two two-story wing rooms with beautifully carved eaves, wooden railings, and Queti beams. The flush gable roof is covered with small gray tiles. The courtyard walls are high and spacious, and the crosswalls are decorated with two types of ornaments: Guanyin gable wall and Yunhuan (a traditional decorative element commonly found on the crosswalls of Chinese traditional architecture. It is a cloud-shaped decoration made of ceramic or wooden materials, usually presenting a flowing and natural form. The purpose of Yunhuan decoration is to enhance the beauty and artistic value of the building, while also carrying symbolic meanings of auspiciousness and good fortune. On the crosswalls, Yunhuan decorations are often combined with other traditional symbols and patterns, creating a unique architectural style and cultural atmosphere), giving them an elegant appearance. In June 2021, it was transformed into the Pudong Hall of Academicians Excellence and the Xinchang Home for Academician.

新场张氏宅第

Zhang Family Residence in Xinchang

年代: 清代
类别: 近现代重要史迹及代表性建筑
保护级别: 浦东新区文物保护单位
利用情况: 居住场所
Era: Qing Dynasty
Category: Modern important historical sites and representative buildings
Conservation level: Pudong New Area district-level protected site
Utilization: Residence

新场张氏宅第原有河西花园和河东住宅二处，现仅存住宅，占地面积 680.22 平方米，建筑面积 1160.14 平方米，坐西朝东，前后四进，砖木结构，硬山灰瓦顶。一进临街门面，原为张信昌绸布庄，有二层楼房 3 间。后为歇山顶砖雕门楼，大门书"京洛传钩，曲江养鸽"联语。

二、三进为前厅和大厅，皆二层楼，左右厢房，有回廊，柳桉木彩色玻璃门窗，石膏线顶，马赛克地坪，具中西合璧特色。四进杂用平房，亦有厢房，房后有马鞍水桥。

The Zhang Family Residence in Xinchang originally consisted of two sections, the Hexi Garden and the Hedong Residence. Currently, only the residence remains, occupying an area of 680.22 square meters with a construction area of 1160.14 square meters. It faces east, with a four-section layout and a brick and wood structure, topped with a flush gable roof and gray tiles. The first section faces the street and used to be the storefront of Zhang Xinchang Silk and Cloth Shop, featuring a two-story building with three rooms. Behind it is a brick-carved entrance gate with a gable and hip roof, with the words "Jing Luo Chuan Gou" and "Qu Jiang Yang Ge" (two tales indicate the Zhangs lived inside) inscribed on the main door.

The second and third sections consist of a front hall and a main hall, both two stories high, with wings on the left and right sides, connected by corridors. The doors and windows are decorated with colored stained glass made of willow and eucalyptus wood, and the ceilings are adorned with gypsum lines. The floors are made of mosaic tiles, combining Chinese and Western styles. The fourth section consists of utility bungalows, and wing rooms, with a saddle-shaped water bridge behind the buildings.

正顺（官）酱园
Zhengshun (official) Sauce and Pickle Shop

年代: 清代
类别: 古建筑
保护级别: 浦东新区文物保护点
利用情况: 居住场所
Era: Qing Dynasty
Category: Ancient architecture
Conservation level: Pudong New Area district-level protected place
Utilization: Residence

　　正顺（官）酱园建于清道光年间。占地面积 1235.58 平方米，建筑面积 989.36 平方米，砖木结构，硬山灰瓦顶，前店后作坊格局。现存房屋 20 间，临街建高围墙，中设石库门，进门为一排店堂，后面是三进坐西朝东的制作工场和露天晒场，后临市河，原有水桥和木桥各 1 座，已毁。是原南汇地区现存的少数古代店铺作坊式建筑之一。该酱园是当时方圆百里规模最大的酱园，售卖货品齐全。全盛时期酱园的员工达 80 余人，酿造的酱油、黄酒、白酒、米醋很受百姓喜爱。

　　The Zhengshun (official) Sauce and Pickle Shop was built during the reign of Daoguang in the Qing Dynasty. It covers an area of 1235.58 square meters, with a building area of 989.36 square meters. The building is a brick and wood structure, with a flush gable roof coverd with gray tiles, and follows a layout with a front shop and a rear workshop. There are currently 20 surviving houses, with a high wall facing the street and a Shikumen in centre. Upon entering, there is a row of shops, followed by three workshops facing east, as well as an open-air drying area. The back of the garden faces the Shihe River, on which there were originally a water bridge and a wooden bridge that have both been destroyed. It is one of the few surviving ancient workshop-style buildings in the former Nanhui area. Zhengshun (official) Sauce and Pickle Shop was once the largest in scale within a hundred li, selling a wide range of goods. During its heyday, the garden employed over 80 people, and its soy sauce, yellow wine, white wine, and rice vinegar were highly popular among the people.

千秋桥
Qianqiu Bridge

年代：清代
类别：古建筑
保护级别：浦东新区文物保护单位
利用情况：开放参观
Era: Qing Dynasty
Category: Ancient architecture
Conservation level: Pudong New Area district-level protected site
Utilization: Open for visit

千秋桥原名仗义桥，是一座单孔石拱桥，清康熙年间由新场人钱纪章集资而建。千秋桥全长 28 米，桥宽 3.9 米，单拱桥洞门高 6 米，东西石阶分别为 21 步和 22 步。桥西两侧，镌有诠释"仗义"桥名的桥联，南侧为"愿天常生好人，愿人常行好事"，北侧为"济人即是济己，种福必须种德"。

The Qianqiu Bridge, originally named Zhangyi Bridge, is a single-arch stone bridge funded and built by Qian Jizhang, a native of Xinchang, during the Kangxi reign of the Qing Dynasty. The Qianqiu Bridge is 28 meters long and 3.9 meters wide, with a single-arch door height of 6 meters. The stone steps on the east and west sides of the bridge are 21 and 22 steps respectively. On the west side of the bridge, there are inscriptions explaining the name "Zhangyi" Bridge. On the south side, it says "May good people always be born, and may people always do good deeds", while on the north side, it says "Helping others is helping oneself, and sowing blessings requires sowing virtues".

新场镇石驳岸及马鞍水桥

Stone Revetment and Saddle-shaped Water Bridge in Xinchang

年代: 明、清
类别: 古建筑
保护级别: 浦东新区文物保护单位
利用情况: 开放参观
Era: Ming and Qing Dynasties
Category: Ancient architecture
Conservation level: Pudong New Area district-level protected site
Utilization: Open for visit

新场镇石驳岸分布于古镇的洪桥港、包桥港、后市河及历史水系沿岸，现存约 1500 米。石驳岸上分布有埠头（当地称"水桥"），现存有明清以来砌筑的马鞍形埠头（马鞍水桥）14 处，部分埠头刻有暗八仙图案。石驳岸及埠头外侧凿有缆船石和下水道，图案丰富且富有美感。石驳岸、埠头、缆船石作为一个整体，是新场古镇江南水乡特色河岸的重要遗存和代表，也是浦东新区境内保存最为完整的传统石驳岸。

The stone revetment in Xinchang Town are distributed along the Hongqiao Port, Baoqiao Port, Houshi River, and the historical water systems of the ancient town. There are about 1500 meters remaining. The revetment is adorned with wharves (locally referred to as "Shuiqiao", i.e. water bridge). There are 14 remaining saddle-shaped wharves (saddle-shaped water bridges), dating back to the Ming and Qing Dynasties. Some of the wharves are engraved with hidden Eight Immortals patterns. Outside the stone revetment and wharves, there are carved cable ship stones and drainage channels, adorned with rich and aesthetically pleasing patterns. The stone revetment, wharves, and cable ship stones, as a whole, are significant remnants and representations of the traditional riverside features of the water town in Xinchang and are the most well-preserved traditional stone revetments within the Pudong New Area.

嘉定镇

Jiading Town

1 嘉定孔庙
嘉定镇南大街 183 号

2 秋霞圃
嘉定镇东大街 314 号

3 法华塔
嘉定镇南大街 349 号

4 嘉定城墙遗址
嘉定镇南大街南城墙公园内；南大街南水关公园内；人民街西首；北大街 271 号嘉定一中附属小学东北侧

5 汇龙潭（应奎山、魁星阁）
嘉定镇塔城路 299 号汇龙潭公园内

6 怡安堂与缀华堂
嘉定镇塔城路 299 号汇龙潭公园内

7 聚善桥
嘉定镇西大街 200 号前

8 德富桥（日晖桥）
嘉定镇法华塔东北侧

1 *Jiading Confucian Temple*
No.183, South Street, Jiading Town

2 *Qiuxia Garden*
No.314, East Street, Jiading Town

3 *Fahua Pagoda*
No. 349, South Street, Jiading Town

4 *Former Site of the Jiading City Wall*
Within the South City Wall Park on South Street; Within the South Water Gate Park on South Street; West end of Renmin Street; No.271 North Street, northeast of Primary School Affiliated to Jiading No.1 High School, Jiading Town

5 *Huilong Pond (Yingkui Mountain, Kuixing Pavilion)*
No.299, Tacheng Road, Huilong Pond Park, Jiading Town

6 *Yi'an Hall and Zhuihua Hall*
No.299, Tacheng Road, Huilong Pond Park, Jiading Town

7 *Jushan Bridge*
In front of No.200, West Street, Jiading Town

8 *Defu Bridge (Rihui Bridge)*
Northeast of Fahua Pagoda, Jiading Town

嘉定城墙遗址（北水关）
Former Site of the Jiading City Wall
(North Water Gate)

嘉定博物馆
Jiading Museum

秋霞圃
Qiuxia Garden

德富桥（日晖桥）
Defu Bridge (Rihui Bridge)

法华塔
Fahua Pagoda

怡安堂
Yi'an Hall

嘉定紫藤园
Jiading Wisteria Garden

汇龙潭公园
Huilong Pond Park

缀华堂
Zhuihua Hall

嘉定孔庙
Jiading Confucian Temple

汇龙潭（应奎山、魁星阁）
Huilong Pond
(Yingkui Mountain,
Kuixing Pavilion)

嘉定城墙遗址（西城墙）
Former Site of the Jiading City Wall
(West City Wall)

嘉定城墙遗址（西水关）
Former Site of the Jiading City Wall
(West Water Gate)

嘉定城墙遗址（南水关）
Former Site of the Jiading City Wall
(South Water Gate)

聚善桥
Jushan Bridge

南城墙公园
South City Wall Park

南水关公园
South Water Gate Park

嘉定城墙遗址（南城墙）
Former Site of the Jiading City Wall
(South City Wall)

嘉行公路 Jiaxing Highway

环城路 Huancheng Road

北大街 Beixia Street

博乐路 Bole Road

温宿路 Wensu Road

清河路 Qinghe Road

城中路 Chengzhong Road

练祁河 Shixin River

塔城路 Tacheng Road

嘉罗公路 Jialuo Highway

塔城路 Tacheng Road

嘉定横沥河 Jiadinghengli River

练祁河 Lianqi River

沪宜公路 Huyi Highway

本图为位置示意，与实际尺寸不符
Illustration is not proportional to the actual scale

嘉定镇
Jiading Town

交通指南：
地铁 11 号线至嘉定北站，换乘公交嘉定 1 路至清河路察院弄站。

Transportation Guide:
Metro Line 11 to Jiading North Station, transfer to Bus Jiading No.1 to Qinghe Road Chayuan Lane Station.

清嘉之土，礼乐之城

嘉定历史悠久，文化底蕴深厚，是江南的一片"清嘉之土，安定之地"。嘉定镇位于嘉定区中部，古为练祁市，自南宋时期嘉定建县以来，长期为嘉定的政治、经济、文化中心。

春秋战国之前，长江口南岸已形成数条古海岸线，在横沥（石岗东侧）、练祁两河交汇之地（今嘉定镇一带）已有先民在此地繁衍生息。南北朝时期，随着人口的迁入和佛教寺庙的兴建，该地在经济、文化等方面得到初步发展。至两宋之际，人口大量迁入，市镇出现，经济得到进一步发展。南宋嘉定十年（1217）嘉定建县，后为"正民风，施教化"，于南宋嘉定十二年（1219）建孔庙。自此，"教化精神"历代延续，为嘉定赢得"教化嘉定"之誉。

明清时期，嘉定的经济、文化建设得到长足的发展，是江南地区重要的商贸集散地和文化重镇，土纱、土布、竹器等大宗交易十分兴盛，尤以土布业最为繁荣。上海开埠后，嘉定人得"西学东渐"风气之先，大力发展实业，1949年以前，嘉定镇商铺星罗棋布，达600多家。同时，嘉定也十分注重文化教育。明清以来，在崇文崇教、忠义爱国传统文化精神的熏陶下，嘉定镇培育出了明代徐学谟、唐时升、孙元化、黄淳耀，清代王敬铭、王鸣盛、秦大成、徐郙，民国顾维钧、吴蕴初、廖世承、胡厥文等重要历史人物。

嘉定镇处于江南水网密集地区，古城区形态近似圆形，曾以城墙与护城河（今环城河）为边界，四面设城门与水关。古城内南北向的横沥河与东西向的练祁河交叉于古镇中心，加上环形的护城河，形成江南古镇中独有的"十字加环"水系格局。

如今，嘉定镇仍保留有部分城墙与水关，镇内划定州桥和西门两片历史文化风貌区，以城中路为界无缝相接。州桥风貌区依托练祁河与横沥河交汇的水系区域成形，在历史上曾是古城的核心区，是嘉定最为繁华的商业街区。西门风貌区以今西大街为核心街巷，是今嘉定镇的发祥地。两风貌区内两宋以来形成的街巷格局保存至今，空间尺度宜人，保留着风格质朴的民居建筑和古老的弹硌路街面。

镇内历史名胜和人文景观众多，"吴中第一"的嘉定孔庙保存完好，历经 800 年风雨的法华塔依旧矗立，秋霞圃、汇龙潭江南园林韵味浓厚，和老街上的古桥、民居一起构成了嘉定丰富而独特的历史文化面貌。嘉定绵延的文脉也凝结出了众多的非物质文化遗产，嘉定竹刻、嘉定锡剧、郁金香酒酿造工艺等颇具地方特色，是彰显古镇文化韵味，唤醒故乡记忆的文化瑰宝。

依托古镇特有的人文禀赋，嘉定镇自 2008 年起每年举办上海孔子文化节等活动，讲好古镇故事，并依托"十字加环"的独特景观，打造"十字河观光带""环城河文化圈"，整体性规划推进历史文化名镇保护传承，使嘉定镇成为了彰显人们故乡记忆的休闲胜地。来到这里，游人可在千步之内漫游宋、元、明、清的古塔、旧庙、名园，观老街风情，品风味小吃，寻找江南古镇的文明印记。

不可移动文物资源

2008 年 10 月，嘉定镇被公布为第四批中国历史文化名镇。镇域内划定西门和州桥两个历史文化风貌保护区，其中西门风貌区面积 44.6 公顷，核心

嘉定镇参观指南 Jiading Town Visiting Guide

游览路线：
嘉定博物馆→秋霞圃→法华塔→嘉定孔庙→汇龙潭公园→嘉定紫藤园→南水关公园→南城墙公园

Tourist Route：
Jiading Museum → Qiuxia Garden → Fahua Pagoda → Jiading Confucian Temple → Huilong Pond Park → Jiading Wisteria Garden → South Water Gate Park → South City Wall Park

古镇美食特产：
小笼包、嘉定竹刻、郁金香酒

Local Specialties：
Xiaolongbao, Jiading Bamboo Carving, Tulip Wine

保护区面积 15.16 公顷；州桥风貌区面积 49.1 公顷，核心保护区面积 15.76 公顷。镇域内现有各级不可移动文物 52 处，包括全国重点文物保护单位 1 处，上海市文物保护单位 3 处，区级文物保护单位 22 处，区文物保护点 26 处，大多位于两个历史文化风貌区内，沿"十字"水脉分布。

A town of purity and stability, worshiping rites and music

Jiading has a long history and profound cultural heritage, known as the "land of purity, place of stability" in Jiangnan. Jiading Town is located in the central part of the Jiading District. In ancient times, it was called Lianqi City. Since Jiading Town was established as a county in the Southern Song Dynasty, it has long been the political, economic, and cultural center of Jiading.

Before the Spring and Autumn Period and the Warring States Period, several ancient coastlines formed on the southern bank of the Yangtze River estuary. In the area where Hengli (east of Shigang) and Lianqi rivers converge (now around Jiading Town), human settlement and reproduction had already taken place. During the period of the Northern and Southern dynasties, with the influx of population and the construction of Buddhist temples, the area experienced initial development in terms of economy and culture. By the time of the Song Dynasty, there was a large influx of population, and market towns emerged, leading to further economic development. In the 10th year of the Southern Song Dynasty's Jiading reign (1217), Jiading County was established, followed by the promotion of "upright moral standards and the spread of education". In the 12th year of the Southern Song Dynasty's Jiading reign (1219), a Confucian temple was built. Since then, the spirit of "education and enlightenment" has been passed down through the generations, earning Jiading the reputation of "Enlightened Jiading".

During the Ming and Qing dynasties, Jiading experienced significant economic and cultural development and became an important commercial and cultural center in the Jiangnan region. It was a bustling hub for trade and commerce, with the trading of goods such as local silk, cotton, bamboo products flourishing. Among them, the local silk industry was particularly prosperous. After Shanghai opened as a port, the people of Jiading embraced the trend of "Western learning for Eastern progress" and vigorously developed industry. Prior to 1949, the town of Jiading was filled with more than 600 shops, showcasing a thriving commercial sector. At the same time, Jiading also placed great importance on cultural education. Since the Ming and Qing dynasties, influenced by the traditional cultural spirit of respect for education, loyalty, righteousness, and patriotism, Jiading cultivated many important historical figures. During the Ming Dynasty, notable figures from Jiading included Xu Xuemo, Tang Shisheng,

Sun Yuanhua, and Huang Chunyao. In the Qing Dynasty, Wang Jingming, Wang Mingsheng, Qin Dacheng, and Xu Fu were prominent individuals from Jiading. In the Republican era, figures like Gu Weijun, Wu Yunchu, Liao Shicheng, and Hu Juewen emerged from Jiading, showcasing their contributions to history and culture.

Jiading Town is located in a densely interconnected water network area in Jiangnan. The layout of the ancient city area is similar to a circle, with the city wall and the moat (now the Huancheng River) serving as boundaries,

and city gates and water gates on all four sides. The north-south Hengli River and the east-west Lianqi River intersect at the center of the ancient town, along with the circular moat, forming a unique "cross plus ring" water system pattern found in ancient towns in Jiangnan.

Nowadays, Jiading Town still retains some city walls and water gates. The town is divided into two historical and cultural style areas: Zhouqiao and Ximen, which seamlessly connect along the Chengzhong Road. The Zhouqiao-style area is formed by the convergence of the

Lianqi River and the Hengli River, and was historically the core area of the ancient city. It is the most prosperous commercial district in Jiading. The Ximen-style area, centered around the present-day West Street, is the birthplace of Jiading Town. The street layout, which has been preserved since the Song Dynasty, still exists in both style areas. It features a pleasant spatial scale and retains simple-style residential buildings and old Tan'ge roads (paved road with gaps, a type of street pavement made from wooden or stone materials, with gaps between the stones or wooden boards, creating a resilient road surface).

There are many historical and cultural attractions in Jiading. The well-preserved Jiading Confucian Temple, known as the "Number One of Wu Region", stands intact. The Fahua Pagoda, which has weathered 800 years of wind and rain, still proudly stands. The beautiful Jiangnan gardens of Qiuxia Garden and Huilong Pond, along with the ancient bridges and residences on the old streets, contribute to the rich and unique historical and cultural landscape of Jiading. Furthermore, Jiading boasts a variety of intangible cultural heritage that reflects its extensive cultural heritage. Jiading bamboo carving, Jiading tin opera, and the craft of brewing tulip liquor all have distinctive local characteristics. These cultural treasures not only showcase the charm of the ancient town's culture but also evoke memories of one's homeland.

Relying on the unique cultural heritage of the ancient town, Jiading Town has been hosting brand events such as the Shanghai Confucius Cultural Festival every year since 2008. These events aim to tell the stories of the ancient town and utilize the unique landscape of the "Cross plus Ring". Through the creation of the "Cross River Sightseeing Belt" and the "Ringed River Cultural Circle", Jiading Town promotes the overall planning and preservation of its historical and cultural landmarks, making it a leisure destination that showcases people's memories of their hometown. Visitors can explore ancient pagodas, temples, and famous gardens from the Song, Yuan, Ming, and Qing dynasties, all within a thousand steps. They can immerse themselves in the charm of the old streets, indulge in local snacks, and discover the cultural imprints of Jiangnan ancient towns.

Immovable Cultural Relics Resources

In October 2008, Jiading Town was announced as the fourth batch of national famous historical and cultural towns. Within the town, two historical and cultural preservation areas were designated: Ximen and Zhouqiao. The Ximen preservation area covers an area of 44.6 hectares, with a core protection area of 15.16 hectares. The Zhouqiao preservation area covers an area of 49.1 hectares, with a core protection area of 15.76 hectares. Currently, there are 52 immovable cultural relics at various levels within the town, including 1 national priority protected site, 3 Shanghai city-level protected sites, 22 district-level protected sites and 26 district-level protected places. Most of these cultural relics are located within the two historical and cultural preservation areas and are distributed along the "cross-shaped" waterways.

嘉定孔庙
Jiading Confucian Temple

年代：宋至清
类别：古建筑
级别：全国重点文物保护单位
利用情况：开放参观
Era: Song to Qing Dynasties
Category: Ancient architecture
Conservation level: National priority protected site
Utilization: Open for visit

嘉定孔庙始建于南宋嘉定十二年（1219），由嘉定县第一任知县高衍孙筹建，于元代重建，有"吴中第一"之称。后经屡次重修、增扩，演变为今日格局。今建筑群占地 10 余亩，合 1.13 万平方米，建筑面积 3380 平方米，是上海地区规格形制保存最为完好的孔庙。中轴线由南至北依次有仰高坊、棂星门、泮池（桥）、大成门、大成殿等建筑。中轴线左有礼门、明伦堂轴线，再东南为当湖书院。庙前方的汇龙潭、应奎山及魁星阁、文昌阁等处，原属孔庙范围，今另辟为汇龙潭公园。1961 年，嘉定博物馆迁设于嘉定孔庙；2005 年，嘉定孔庙辟建为上海中国科举博物馆。

Jiading Confucian Temple, also known as Jiading Kong Miao, was originally built in the 12th year of the Jiaqing reign of the Southern Song Dynasty (1219) by Gao Yansun, the first county magistrate of Jiading. During the Yuan Dynasty, it was rebuilt and earned the title of "Number One of Wu Region". After several reconstructions and expansions, it evolved into its present-day layout. The temple complex covers an area of over 10 mu, totaling 11 300 square meters, with a building area of 3380 square meters. It is the best-preserved Confucian Temple in the Shanghai region in terms of its size and architectural form. The central axis of the complex includes buildings such as the Yanggao Archway, Lingxing Gate, Pan Pond (Bridge), Dacheng Gate, and Dacheng Hall, arranged from south to north. To the left of the central axis, there is the Limen and a line leading to the Minglun Hall, while the southeast area is occupied by the Danghu Academy. In front of the temple, the Huilong Pond, Yingkui Mountain, Kuixing Pavilion, and Wenchang Pavilion, originally part of the temple, have been developed separately into the Huilong Pond Park. In 1961, the Jiading Museum was established within the Confucian Temple and in 2005, it was converted into the Shanghai China Imperial Examination Museum.

秋霞圃
Qiuxia Garden

年代: 明代
类别: 古建筑
级别: 上海市文物保护单位
利用情况: 开放参观
Era: Ming Dynasty
Category: Ancient architecture
Conservation level: Shanghai city-level protected site
Utilization: Open for visit

秋霞圃由始建于明中期的龚氏园、金氏园，明后期的沈氏园，以及明初移建于此的城隍庙合并而成，清初，龚氏园改名秋霞圃。清中期，龚氏园、沈氏园归属城隍庙（邑庙），为后园。咸丰十年（1860），园景毁于战事。光绪年间（1875—1908）陆续重建。民国时期陆续修缮与新建。民国三十五年（1946）改称为"邑庙公园"。1960年恢复"秋霞圃"名。秋霞圃现占地面积3.28公顷。园内有亭台楼阁、池塘山石，以及古树名木、翠竹花卉等，移步换景，秀色宜人，是市民和中外游客休闲旅游的极佳去处。

Qiuxia Garden is a park formed by the merging of several gardens: Gong Family Garden and Jin Family Garden, both established in the mid-Ming Dynasty; Shen Family Garden, constructed in the late Ming Dynasty; and the Chenghuang Temple, which was relocated here in the early Ming Dynasty. In the early Qing Dynasty, Gong Family Garden was renamed Qiuxia Garden. In the mid-Qing Dynasty, Gong Family Garden and Shen Family Garden became part of the Chenghuang Temple (Yi Miao), serving as its back garden. In the 10th year of the Xianfeng reign (1860), the garden was destroyed during warfare, and then gradually rebuilt during the Guangxu reign (1875–1908). Additional renovations and new constructions were carried out during the Republican era. In the 35th year of the Republic of China (1946), it was renamed "Yi Miao Park". In 1960, it was restored and given the name "Qiuxia Garden". The park now covers an area of 3.28 hectares. Within the premises there are pavilions, pagodas, ponds, rocks, ancient trees, beautiful flowers and lush bamboo groves. Exploring the park allows visitors to enjoy ever-changing scenery, making it a delightful destination for leisure and tourism for both locals and tourists from China and abroad.

法华塔
Fahua Pagoda

年代: 明代
类别: 古建筑
级别: 上海市文物保护单位
利用情况: 开放参观
Era: Ming Dynasty
Category: Ancient architecture
Conservation level: Shanghai city-level protected site
Utilization: Open for visit

法华塔，又名金沙塔，始建于南宋开禧年间（1205—1207），明万历三十六年（1608）重建。通高41米，砖木结构，四面七层楼阁式，石砌台基。塔刹由铁铸仰覆莲、相轮和宝瓶构成。每层挑出斗拱腰檐和平座，四周设围廊，层间两面设上下进出口门洞，南北、东西对应，层层交错，门内上方设藻井，塔内置木梯。塔身每层四面及塔心室分别设有龛室。法华塔为嘉定古城地标性建筑，昔日登上塔顶，全城景致一览无遗。

The Fahua Pagoda (also known as Jinsha Pagoda) was originally built during the Kaixi reign of the Southern Song Dynasty (1205–1207) and was rebuilt in the 36th year of the Wanli reign of the Ming Dynasty (1608). It has a total height of 41 meters and is built with brick and wood structure. It has a four-sided, seven-story tower with a stone base. The top of the pagoda is formed by an inverted lotus, a wheel of dharma, and a treasure vase, all made of iron. Each level of the pagoda features Dougong (bucket arch), cornices, and flat platform, with surrounding corridors. There are entrance and exit doorways on both sides of each level, corresponding to each other in the north-south and east-west directions, creating an interlocking pattern. Above the doors, there are ornate decorative recesses, and wooden stairs inside the pagoda. Each level of the tower has shrine rooms on all four sides, as well as a central chamber. The Fahua Pagoda is a landmark building in the ancient city of Jiading. In the past, climbing to the top of the pagoda allowed a panoramic view of the entire city.

嘉定城墙遗址
Former Site of the Jiading City Wall

年代：宋至清
类别：古建筑
级别：上海市文物保护单位
利用情况：开放参观
Era: Song to Qing Dynasties
Category: Ancient architecture
Conservation level: Shanghai city-level protected site
Utilization: Open for visit

嘉定城墙始建于南宋嘉定十二年（1219），有城壕（护城河），设东西南北四关。元至正十八年（1358），加固土城，四关用砖石建城门，并设水门（水关）。明代多次增筑，为防御倭寇侵犯发挥了重要作用。古城平面略呈圆形，周长2266丈余，有东、西、南、北四座城门，东、西、南、北四水关，以及内堑、外堑（即内外护城河）。在江南地区，像嘉定这样坚固宏伟的县城墙实属鲜见，时人称其规模，谓"东南诸城称雄嘉定矣"。民国后，城墙城门陆续被拆毁，内城河被填埋，现仅剩南城墙、南水关、西城墙、西水关、北水关遗址五处。

The Jiading City Wall, built in the 12th year of the Jiading reign of the Southern Song Dynasty (1219), featured moats and four gates in the east, west, south, and north directions. In the 18th year of the Zhizheng reign of the Yuan Dynasty (1358), the earth wall was reinforced, and the gates were rebuilt using bricks and stones. Water gates were also established. During the Ming Dynasty, the city wall was expanded multiple times, playing an important role in defending against Japanese pirate invasions. The layout of the ancient city is somewhat circular, with a perimeter of over 2266 zhang (a traditional Chinese unit of length, (1 zhang is approximately 3.33 meters). It has four city gates in the east, west, south, and north directions, four water gates, as well as inner and outer moats (also known as inner and outer defensive ditches). In the Jiangnan region, it is rare to find a solid and magnificent county city wall like Jiading's. People at that time praised its scale, calling it the "paramount city among the cities in the southeast". After the Republic of China, the city walls and city gates were gradually demolished, and the inner moat was filled. Currently, only remnants of the south city wall, south water gate, west city wall, west water gate, and north water gate remain visible.

汇龙潭（应奎山、魁星阁）

Huilong Pond (Yingkui Mountain, Kuixing Pavilion)

年代：明、清
类别：古建筑
级别：嘉定区文物保护点
利用情况：开放参观
Era: Ming and Qing Dynasties
Category: Ancient architecture
Conservation level: Jiading district-level protected place
Utilization: Open for visit

汇龙潭包括应奎山、魁星阁两处。汇龙潭开凿于明万历年间，时有五条河流于此交汇，应奎山凸立潭南，潭东岸建魁星阁，曾为嘉定孔庙建筑群的一部分。民国十八年（1929）同孔庙一起辟为公园，名"奎山公园"。

　　应奎山堆建于明天顺四年（1460）。当年重建孔庙大成殿时，以殿南首留光寺与孔庙"相望而不相类"，遂在寺与庙之间筑山以障，取名应奎。魁星阁原名魁星亭，始建于明万历三十一年（1603），清道光七年（1827）改建为阁，1976年重建。方形二层，四角攒尖顶。四面设门，分别题有"苍龙""白虎""朱雀""玄武"四额。

Huilong Pond consists of two locations, Yingkui Mountain and Kuixing Pavilion. Huilong Pond was excavated during the Wanli reign of the Ming Dynasty, where five rivers converged. Yingkui Mountain stood prominently to the south of the pool, and Kuixing Pavilion was built on the east bank of the pool. It was once part of the architectural complex of the Confucian Temple in Jiading. In the 18th year of the Republic of China (1929), it was converted into a park together with the Confucian Temple and named "Kuishan Park".

Yingkui Mountain was constructed in the 4th year of the Tianshun reign of the Ming Dynasty (1460). During the reconstruction of the Dacheng Hall of the Confucian Temple, the Liuguang Temple to the south of the hall was considered to "positions are relative but actually belong to different categories". As a result, a mountain was erected between the temple and the hall to obstruct the view, and it was named Yingkui. Kuixing Pavilion was initially constructed in the 31st year of the Wanli reign of the Ming Dynasty (1603). It was renovated and transformed into a pavilion in the 7th year of the Daoguang reign of the Qing Dynasty (1827) and underwent reconstruction in 1976. It is a square, two-story building with a pointed roof at each corner. Each side of the pavilion features a door inscribed with the words "Canglong" "Baihu" "Zhuque" and "Xuanwu".

怡安堂与缀华堂
Yi'an Hall and Zhuihua Hall

年代：清代
类别：古建筑
级别：嘉定区文物保护点
利用情况：开放参观
Era: Qing Dynasty
Category: Ancient architecture
Conservation level: Jiading district-level protected place
Utilization: Open for visit

　　怡安堂与缀华堂原位于嘉定镇人民街，由廖寿丰、廖寿恒兄弟俩在清光绪年间购置的明代礼部尚书徐学谟"世忠堂"遗址改建而成。旧时规模宏大，有楼房56间，为嘉定名宅显第之冠，俗称"廖家大院"。怡安堂为"廖家大院"正厅，缀华堂为廖家读书、作文、会友之处。两者均坐西北朝东南，砖木结构，面宽三间，硬山式小青瓦屋面，于 20 世纪 80 年代迁入汇龙潭公园。

The Yi'an Hall and Zhuihua Hall were originally located on Renmin Street in Jiading Town. They were rebuilt from the site of the "Shizhong Hall" of Xu Xuemo, an official in the Ming Dynasty. The hall was purchased by the brothers Liao Shoufeng and Liao Shouheng, during the Guangxu period of the Qing Dynasty. The grand complex consisted of 56 buildings and was commonly known as the "Liao Family Residence", which was regarded as the pinnacle of Jiading's prestigious residences. The Yi'an Hall served as the main hall of the "Liao Family Residence" while the Zhuihua Hall was used for activities such as reading, writing, and socializing by the Liao family. Both of the two halls faces southeast, having brick and wood structure, three bays, and flush gable roof covered with small gray tiles. Both halls were relocated to Huilong Pond Park in the 1980s.

聚善桥
Jushan Bridge

年代：明代
类别：古建筑
级别：嘉定区文物保护单位
利用情况：开放参观
Era: Ming Dynasty
Category: Ancient architecture
Conservation level: Jiading district-level protected site
Utilization: Open for visit

聚善桥，俗名虬桥，因建桥捐助者多为女性，故又俗称"女桥"。始建于明洪武十三年（1380），万历四十年（1612）重建。清同治三年（1864）重修，后又于1990年重修。南北走向，跨练祁河。单孔石拱桥，桥身长近28米，宽4米余，桥孔净跨9.30米，拱高5米，是嘉定城区及其周边现存跨度最大的一座古代石拱桥。

The Jushan Bridge, commonly known as the Qiu Bridge, is also called the "Women's Bridge" because many women contributed to its construction. It was initially built in the 13th year of the Hongwu reign of the Ming Dynasty (1380), reconstructed in the 40th year of the Wanli reign (1612), and repaired again in the 3rd year of the Tongzhi reign of the Qing Dynasty (1864). It was later renovated in 1990. The bridge, which spans the Lianqi River, runs in a north-south direction. It is a single-arch stone bridge with a length of nearly 28 meters and a width of over 4 meters. The clear span of the bridge is 9.30 meters, with an arch height of 5 meters. It is the largest existing ancient stone arch bridge in the Jiading urban area and its surrounding areas.

德富桥（日晖桥）
Defu Bridge (Rihui Bridge)

年代：清代
类别：古建筑
级别：嘉定区文物保护单位
利用情况：开放参观

Era: Qing Dynasty
Category: Ancient architecture
Conservation level: Jiading district-level protected site
Utilization: Open for visit

德富桥又称日晖桥，俗称石灰桥，始建于明成化七年（1471），清光绪十一年（1885）重建。东西走向，跨南横沥北口，是一座单孔石拱桥。桥身长18米余，宽3.5米，桥孔净跨8.4米，拱高4米，南北两侧桥联石分别刻"迎潭水南来涵濡圣泽，障娄潮东去容与中流"和"一江澄练塔影虹垂，四面回澜冈身龙卧"。桥下水面可见法华塔倒影，"金沙夕照"成为嘉定胜景之一（金沙即法华塔别名）。

Defu Bridge, also known as Rihui Bridge and commonly called Shihui Bridge, was first built in the 7th year of the Chenghua reign of the Ming Dynasty (1471) and later rebuilt in the 11th year of the Guangxu reign of the Qing Dynasty (1885). It spans across the mouth of South Hengli Canal, running from east to west. It is a single-arch stone bridge, measuring over 18 meters in length and 3.5 meters in width, with a clear span of 8.4 meters and an arch height of 4 meters. On the two sides of the bridge, there are inscriptions that read "The water of Huilong Pond comes from the south, nourished by the grace of Confucius, the tide of the Lou (Liu) River comes from the north and intersects with the Lianqi River" and "The reflection of the pagoda casts a rainbow on the clear river, the dragon lies dormant on the hills surrounding it from all directions". Underneath the bridge, the reflection of the Fahua Pagoda can be seen on the water surface, and the "Jinsha Sunset" has become one of Jiading's scenic sights. Jinsha is another name for the Fahua Pagoda.

南翔镇

Nanxiang Town

1 南翔寺砖塔
南翔镇解放街香花桥北堍

2 古猗园
南翔镇沪宜公路 218 号

3 尊胜陀罗尼经幢
南翔镇沪宜公路 218 号古猗园内

4 天恩桥
南翔镇沪宜公路辅路与嘉好路交叉口东南侧

5 纤桥
南翔镇天恩桥西南侧

6 鹤槎山
南翔镇沪宜公路真南路东北侧

1 *Nanxiang Temple Brick Pagoda*
The north end of Xianghua Bridge, Jiefang Street, Nanxiang Town

2 *Guyi Garden*
No.218, Huyi Highway, Nanxiang Town

3 *Zunsheng Dharani Sutra Pillars*
No.218, Huyi Road, located inside Guyi Garden, Nanxiang Town

4 *Tian'en Bridge*
Southeast side of the intersection between the auxiliary road of Huyi Highway and Jiahao Road, Nanxiang Town

5 *Qian Bridge*
Southwest of Tian'en Bridge, Nanxiang Town

6 *Hecha Mountain*
Northeast side of Zhen'nan Road, Huyi Highway, Nanxiang Town

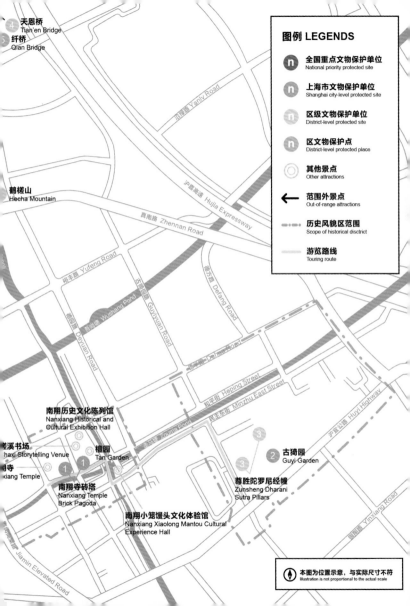

图例 LEGENDS

全国重点文物保护单位
National priority protected site

上海市文物保护单位
Shanghai city-level protected site

区级文物保护单位
District-level protected site

区文物保护点
District-level protected place

其他景点
Other attractions

范围外景点
Out-of-range attractions

历史风貌区范围
Scope of historical district

游览路线
Touring route

天恩桥
Tian'en Bridge

纤桥
Qian Bridge

鹤槎山
Hecha Mountain

沪嘉高速 Hujia Expressway

真南路 Zhennan Road

裕丰路 Yufeng Road

将丰路

古猗园路 Guyiyuan Road

乌漳浜 Wushang Pond

德园路 Deyuan Road

德芳路 Defang Road

和平街 Heping Street

民主东街 Minzhu East Street

沪宜公路 Huyi Highway

走马塘 Zoumu Pond

南翔历史文化陈列馆
Nanxiang Historical and Cultural Exhibition Hall

槎溪书场
haxi Storytelling Venue

檀园
Tan Garden

iang Temple

南翔寺砖塔
Nanxiang Temple Brick Pagoda

南翔小笼馒头文化体验馆
Nanxiang Xiaolong Mantou Cultural Experience Hall

尊胜陀罗尼经幢
Zunsheng Dharani Sutra Pillars

古猗园
Guyi Garden

银翔路 Yinxiang Road

Jiamin Elevated Road

本图为位置示意，与实际尺寸不符
Illustration is not proportional to the actual scale

南翔镇

Nanxiang Town

交通指南：
地铁 11 号线至南翔站，步行或换乘公交 62 路至云翔寺站。

Transportation Guide:
Metro Line 11 to Nanxiang Station, walk or transfer to Bus No.62 and get off at Yunxiang Temple Station.

东南都会，文荟槎溪

南翔镇位于嘉定区东南部，以历史悠久、风貌宜人、市井繁华、民俗和美食丰富而著称。

南翔，因境内有上、中、下三条槎浦，故又名"槎溪"。据地方志载，南朝梁天监四年（505），白鹤南翔寺建于此，后因寺成镇，遂以寺得名。唐代中叶，这里已是富庶的鱼米之乡。至宋元时期，为东南巨镇。明初，经济繁荣程度已为全县各市镇之首。嘉靖年间，屡遭倭寇焚掠，经济一度衰退。至隆庆、万历年间，逐渐复兴。

明清时期为土布业集散中心，徽商侨寓，百货俱集，甲于诸镇，有"银南翔"之称。近代以来，随着铁路、公路穿镇境而过，市况更盛，民族工业也随之兴起。这里人文荟萃，明清两代有进士、举人、贡生近百名。文人雅士竞相兴建宅第园林，曾建园林20多座，故有"小小南翔赛苏城"之誉。

千年的历史积淀，为南翔留下了丰富的文化遗产。古镇留存有唐宋以来形成的"十"字水系，镇中为十字港、横沥、上槎浦、走马塘、封家浜四条河道交汇于镇中心的太平桥南。除了四大干河以外，镇的周围东南西北各

有河湾：东为五圣庙湾，西为侯家湾，南为薛家湾，北为鹤颈湾，似佛教中的"卍"字状。明清以来，南翔镇形成"东西五里，南北三里"的市镇范围，以及"以塔为心""东园西寺"的格局，保存至今的五代南翔双塔位于镇中心，明代以来营造的古猗园位于镇之东南。镇内小桥流水，粉墙黛瓦，屋舍相连，商铺栉比，古典园林风景宜人，寺观塔幢法相庄严，呈现出典型的江南地区市镇风貌。

除了古迹众多，南翔还有历史文化陈列馆、小笼馒头文化体验馆等一批展馆，可深度体验。每年还有古猗园竹文化艺术节、南翔小笼文化展、南翔戏曲庙会等活动，非物质文化遗产、文物古迹与传统文化相结合。漫步南翔，走一走古镇特色的弹硌路，尝一尝老味道的特色小吃，看一看江南风韵的古桥民居，可以感受到千年古镇的历史气息和如诗的生活。

不可移动文物资源

2010 年 7 月，南翔镇被公布为第五批中国历史文化名镇，镇域内有南翔双塔和古猗园两个历史文化风貌区，总面积 68.82 公顷，其中双塔历史文化风貌保护区面积 20.9 公顷，核心保护区面积 6.73 公顷，古猗园历史文化风

貌保护区面积 47.92 公顷，核心保护区面积 9.95 公顷。镇域内现有各级不可移动文物 26 处，包括上海市文物保护单位 2 处，区级文物保护单位 8 处，区文物保护点 16 处。古镇文物荟萃，有唐代的经幢、五代时期的双塔、明代的古猗园以及明清以来的民居、商铺、桥梁等文物古迹。

A town like a metropolis, rich in cultural diversity

Nanxiang Town is located in the southeastern part of Jiading District and is known for its long history, pleasant scenery, bustling marketplaces, rich folk customs, and delicious food.

Nanxiang is also known as "Cha Xi" because there are three branches of the Cha River in the area. According to local records, the Baihe Nanxiang Temple was built here in the 4th year of the Tianjian reign of the Southern Liang Dynasty (505), and the town took its name from the completion of the temple. By the middle of the Tang Dynasty, it had become a prosperous area known for its agriculture and fisheries. During the Song and Yuan dynasties, it was a major town in the southeast. In the early Ming Dynasty, its economic prosperity surpassed that of other towns in the county. However, during the Jiajing period, it suffered frequent attacks and plunder by Japanese pirates, leading to a temporary economic decline. It gradually revived during the Longqing and Wanli periods. During the Ming and Qing dynasties, it became a hub for the linen industry, attracting Huizhou merchants and a wide variety of commodities. This made it surpass other towns and earning it the title of "Silver Nanxiang". In modern times, with the construction of railways and highways passing through the town, its prosperity has further increased, and ethnic industries have also flourished. The area has a rich cultural heritage with nearly a hundred successful examinees in the imperial examinations during the Ming and Qing dynasties. Literati competed to build their own gardens, and there used to be more than 20 gardens in Nanxiang, earning it the reputation of "Nanxiang, a small city rivaling Suzhou".

With a history spanning thousands of years, Nanxiang has inherited a rich cultural heritage. The ancient town still maintains the "cross" water system originated in the Tang and Song dynasties. At the center of the town lies Shizi Harbor, where the four rivers of Hengli, Shangchapu, Zoumatang and Fengjiabang converge at the southern end of Taiping Bridge. In addition to the four main rivers, there are also river bays surrounding the town in the east, west, south and north: Wushengmiao Bay in the east, Houjiawan Bay in the west, Xuejiawan Bay in the south, and Hejingwan Bay in the north, resembling the shape of the " 卍 "(swastika) in Buddhism. Since the Ming and Qing dynasties, Nanxiang has defined its market town area as "five li east-west, three li north-south", with a layout featuring of "tower as the heart" and "east garden, west temple". The ancient twin towers of Nanxiang, built during the Five Dynasties period, are located in the center of the town, while the Guyi Garden built since the Ming Dynasty, is situated in the southeast of the town. The town is adorned with small bridges and flowing water, along with white walls and black tiles. The inter-connected buildings and well-organized shops present a captivating classical garden scenery, while the temples and towers emanate a solemn atmosphere, showcasing the typical appearance of a market town in the Jiangnan region.

In addition to numerous historical sites, Nanxiang also has a number of exhibition halls such as the Historical and Cultural Exhibition Hall and the Xiaolong Mantou Cultural Experience Hall, where visitors can have a deep cultural experience. Every year, there are also events such as the Guyi Garden Bamboo

Culture and Art Festival, Nanxiang Xiaolong Culture Exhibition, and Nanxiang Opera Temple Fair, showcasing the combination of intangible cultural heritage, cultural relics, and traditional culture. Taking a stroll in Nanxiang, walking along the characteristic Tan'ge roads, tasting the local snacks with their unique flavors, and admiring the ancient bridges and residential buildings with the charm of the Jiangnan region, one can feel the historical atmosphere of this thousand-year-old town and its poetic way of life.

Immovable Cultural Relics Resources

In July 2010, Nanxiang Town was announced as the fifth batch of national famous historical and cultural towns. Within the town's jurisdiction, there are two historical and cultural areas, namely, the Nanxiang Twin Towers and Guyi Garden, with a total area of 68.82 hectares. Among them, the historical and cultural protection area of the Twin Towers covers 20.9 hectares, with a core protection area of 6.73 hectares. The historical and cultural protection area of Guyi Garden covers 47.92 hectares, with a core protection area of 9.95 hectares. There are currently 26 immovable cultural relics at various levels within the town, including 2 Shanghai city-level protected sites, 8 district-level protected sites and 16 district-level protected places. The ancient town is rich in cultural relics, including Tang Dynasty sutra pillars, twin towers from the Five Dynasties period, Guyi Garden from the Ming Dynasty, as well as ancient residential buildings, shops, bridges, and other cultural relics and historical sites from the Ming and Qing dynasties.

南翔寺砖塔
Nanxiang Temple Brick Pagoda

年代：五代至北宋
类别：古建筑
保护级别：上海市文物保护单位
利用情况：开放参观
Era: Five Dynasties to Northern Song Dynasty
Category: Ancient architecture
Conservation level: Shanghai city-level protected site
Utilization: Open for visit

南翔寺砖塔，共两座，俗称南翔寺双塔，建于五代至北宋初年，形制相同，底层直径为 1.86 米、通高 11 米，灰砖砌筑，仿木结构楼阁式，七层八面，面积各约 4 平方米。每级四面为壶门，四面为直棂窗，设腰檐、平座、栏板，檐下施五铺作单杪单昂斗拱。八角形攒尖灰瓦顶，顶上立相轮、刹杆、宝珠构成的铁铸塔刹。南翔寺砖塔是南翔千年古镇的标志，旧时为南翔一景，沐浴在朝露暮霭之中，称"双塔晴霞"。塔上火焰形壶门、简朴的直棂窗、精巧的斗拱、细腻的栏板和挺秀的塔刹，无不显示出其建筑工艺的精美。

Nanxiang Temple Brick Pagoda, also known as Nanxiang Temple Twin Pagodas, consists of two pagodas. They were built during the Five Dynasties and early Northern Song Dynasty. The pagodas have the same architectural style, with a bottom diameter of 1.86 meters and a total height of 11 meters. They are constructed with gray bricks, imitating the style of wooden tower. Each pagoda has seven floors and eight sides, with an area of approximately 4 square meters each. Each level has arched doorways on four sides and straight lattice windows on the other four sides, with waist eaves, flat platforms, and railings. Under the eaves (there) are Dougong (bucket arch) components. The pagodas are topped with an octagonal gable roof made of gray tiles, on which stands an iron cast pagoda spire composed of a wheel, a mast, and a treasure bead. Nanxiang Temple Brick Pagoda is a symbol of the thousand-year-old town of Nanxiang and was once a famous sight in Nanxiang. It radiates a magical charm in the morning dew and evening haze, earning it the name "Twin Pagodas in Clear Sunset". The flame-shaped arched doorways, simple lattice windows, exquisite Dougong (bucket arch), delicate railings, and elegant pagoda spires on the pagodas all showcase the exquisite craftsmanship of its architecture.

古猗园
Guyi Garden

年代： 明代
类别： 古建筑
保护级别： 上海市文物保护单位
利用情况： 开放参观
Era: Ming Dynasty
Category: Ancient architecture
Conservation level: Shanghai city-level protected site
Utilization: Open for visit

古猗园，始建于明嘉靖年间（一说万历年间），为河南通判闵士籍的私家花园，初名"借园"，后以《诗经》"绿竹猗猗"句，更名猗园，由朱稚征精心设计布置，有"十亩之园，五亩之宅"之说。园内遍植数十种名贵竹种，筑亭台楼阁其间，立柱、椽子、长廊上无不刻着千姿百态的竹景，形成独特的造园艺术。明末归贡生李宜之，后又归陆、李两姓。清乾隆十一年（1746）为苏州洞庭山人叶锦购得，拓地增筑 30 余处景点，更名"古猗园"，相沿至今。乾隆五十三年（1788）捐作城隍祠。民国时期，部分园景毁于战火。1946 年之后，部分重建，并新建微音阁。以戏鹅池为中心，主要建筑有逸野堂、南厅、浮筠阁等。古猗园几经扩建，现占地面积近 10 万平方米。

Guyi Garden, built during the Ming Dynasty in the Jiajing reign (some say it was during the Wanli era), was originally a private garden of Min Shiji, a judge in Henan. Originally named Jie Garden, it was later renamed Yi Garden [" 猗猗 "(yiyi) means it is graceful], derived from the line "green bamboo sways gracefully" from *The Book of Songs* , and it was meticulously designed and arranged by Zhu Zhizheng. It is often referred to as "a garden of ten mu with a residence of five mu". The garden is filled with dozens of rare bamboo species, and pavilions, terraces, and buildings are interspersed among them. The columns, beams, and long corridors are all carved with various bamboo scenes, forming a unique garden art. In the late Ming Dynasty, it was owned by Li Yizhi, a successful scholar, and later passed down to the Lu and Li families. In the 11th year of the Qianlong period of the Qing Dynasty (1746), it was purchased by Ye Jin, a prominent figure in Suzhou, who expanded and added more than 30 scenic spots, renaming it "Guyi Garden", a name that continues to this day. In the 53rd year of Qianlong (1788), it was donated to be used as the Chenghuang Temple. During the period of the Republic of China, most of the garden was destroyed in the war. After 1946, it was partially rebuilt, and the Weiyin Pavilion was newly constructed. Centered around the Xi'e Pond, the main buildings include the Yiye Hall, South Hall, and Fuyun Pavilion. Guyi Garden has been expanded several times and currently covers an area of nearly 100 000 square meters.

尊胜陀罗尼经幢
Zunsheng Dharani Sutra Pillars

年代：唐代
类别：古建筑
保护级别：嘉定区文物保护单位
利用情况：开放参观
Era: Tang Dynasty
Category: Ancient architecture
Conservation level: Jiading district-level protected site
Utilization: Open for visit

尊胜陀罗尼经幢,共二座,形制相同,为南翔寺旧物,对称立于大雄宝殿之前,里人莫少卿捐建,分别建于唐咸通八年(867)、乾符二年(875),原均七级八面,仰莲基座,造型壮丽挺秀,镌刻精美。幢身镌有尊胜陀罗尼经文及卷云、莲瓣等纹饰,幢顶镌有狮首和四天王像。北宋太平兴国五年(980)、元元统二年(1334)、清嘉庆中叶先后重修。清末寺废,解放时已倒塌。1959年按原样移至南翔镇古猗园。一座在古猗园内南厅北侧,为唐咸通八年建,青石质,目前存八角形基座、幢身和宝顶下部。幢身剥蚀严重,顶上宝刹于1969年遭雷击毁。幢身有宋治平二年(1065)及清嘉庆年间南翔寺僧及发愿人维修经幢的题记。一座在古猗园内微音阁南侧,为唐乾符二年建,高十余米。两座经幢出于盛唐工艺,造型秀丽、雕凿精美,人物形态丰腴典雅、雍容自若,各种纹饰简洁传神、装饰性强,是典型唐代雕饰风格。

The Zunsheng Dharani Sutra Pillars, consisting of two pillars, are ancient artifacts of Nanxiang Temple. They stand symmetrically in front of the Mahavira Hall. They were donated and built by a local person named Mo Shaoqing in the 8th year of the Xiantong reign of the Tang Dynasty (867) and the 2nd year of the Qianfu reign of the Tang Dynasty (875). Originally, they both had seven levels and eight sides with a lotus-shaped base, and they had magnificent and elegant designs with exquisite carvings. The body of the pillars is engraved with the Zunsheng Dharani sutra text and various decorative patterns such as rolling clouds and lotus petals, while the top of the pillars is adorned with lion heads and images of the Four Heavenly Kings. They underwent repairs during the 5th year of the Taiping Xingguo reign of the Northern Song Dynasty (980), the 2nd year of the Yuantong reign of the Yuan Dynasty (1334), and the middle period of the Jiaqing reign of the Qing Dynasty. The temple was abandoned in the late Qing Dynasty, and the pillars had collapsed by the time of the liberation. In 1959, they were moved to Guyi Garden in Nanxiang Town. One of the pillars, which was built in the 8th year of the Xiantong reign of the Tang Dynasty and made of bluestone, is currently preserved with its octagonal base, pillar body, and the lower part of the precious top. The pillar body has suffered severe weathering, and the precious top was destroyed by lightning in 1969. There are inscriptions on the pillar body from the 2nd year of the Zhiping reign of the Song Dynasty (1065) and the Jiaqing period of the Qing Dynasty, recording the repairs made by monks and vow makers from Nanxiang Temple. The other pillar, located on the south side of the Weiyin Pavilion in Guyi Garden, was built in the 2nd year of the Qianfu reign of the Tang Dynasty and is over ten meters tall. Both pillars showcase exquisite craftsmanship of the flourishing Tang Dynasty, featuring beautiful and finely carved designs, elegant and graceful figures, and simple yet expressive decorative patterns. They represent the typical carving style of the Tang Dynasty.

天恩桥
Tian'en Bridge

年代: 清代
类别: 古建筑
保护级别: 嘉定区文物保护单位
利用情况: 开放参观
Era: Qing Dynasty
Category: Ancient architecture
Conservation level: Jiading district-level protected site
Utilization: Open for visit

天恩桥，又名真圣堂桥，始建年代不详，清顺治年间（1644—1661）重建，易今名。清雍正、乾隆年间先后重修。同治十三年（1874）重建。该桥为三孔石拱桥，东西向，跨横沥河。石阶桥面，通长近40米，顶部宽3.5米，中孔净跨11米余，两边桥孔净跨5.9米。桥孔两侧刻楹联，南为"云际龙飞高凌百尺，波间虹卧彩耀三槎""境接吴淞势挟汪洋通万顷，名颜真圣义兼廉让媲千秋"。北为："行看桂子月中落，定有仙槎海上来""人杰地灵白鹤来飞传胜迹，风恬浪静彩虹遥映镇槎溪"。桥身西南侧嵌同治十三年"饬修监修官绅衔名"碑一方，护栏刻桥名及铭记。整座桥气势雄伟又不失秀丽灵气。

在西桥墩，建有贴水平桥一座，作为纤夫的专用通道，是现在立交桥的雏形。天恩桥以建造技术高超、造型优美、人文资源丰富闻名于世，有"嘉定第一桥"的美誉。

Tian'en Bridge, also known as Zhenshengtang Bridge, was originally built in an unknown period and was reconstructed during the Shunzhi reign of the Qing Dynasty (1644–1661). It was later renamed Tian'en Bridge. During the reigns of Yongzheng and Qianlong in the Qing Dynasty, further renovations were made to the bridge. In the 13th year of the Tongzhi reign (1874), the bridge was rebuilt once again. The bridge is a three-arch stone bridge, spanning the Hengli River in an east-west direction. The stone steps on the bridge are nearly 40 meters long, with a width of 3.5 meters at the top. The central arch spans more than 11 meters, while the two side arches span 5.9 meters each. Inscriptions on each side of the bridge display couplets. On the south side, the inscription reads "Dragons soar high in the clouds, rainbows shine amidst the waves; Wujiang's grandeur connects with big oceans, its reputation as a true sacred place lasts for eternity". On the north side, it reads "Viewing the moon descending behind the osmanthus trees, surely there is a divine ship coming from the sea; With outstanding individuals and the spirit of the land, white cranes fly and spread the tales of the flourishing town reflected by the peaceful waves". On the southwest side of the bridge, a stone tablet is engraved with the names of the officials involved in the reconstruction in the 13th year of the Tongzhi reign, as well as the name of the bridge and a commemoration. The entire bridge exemplifies grandeur and beauty while retaining an air of elegance and charm.

On the western bridge pier, there is a horizontal bridge attached to it, serving as a dedicated passage for boat trackers, and it is a prototype of the current overpass. Tian'en Bridge is renowned for its superb craftsmanship, beautiful scenery, and abundant cultural connotation, which has earned it the reputation of being the "Number One Bridge in Jiading".

纤桥
Qian Bridge

年代: 近现代
类别: 近现代重要史迹及代表性建筑
保护级别: 嘉定区文物保护点
利用情况: 开放参观
Era: Modern times
Category: Modern important historical sites and representative buildings
Conservation level: Jiading district-level protected place
Utilization: Open for visit

　　纤桥,始建于清代,民国时期重建。该桥为三架平梁桥,花岗石材质,通长18.7米,宽1.6米,中跨7米。桥面采用长方形条石平铺,桥柱为并列式长方形条石,两端石砌桥墩,为嘉定横沥河沿线唯一存在的主要供过往船只纤夫通行的石板桥。

　　横沥河南接吴淞江,北通浏河,水运繁忙,历来为嘉定、太仓等地通向上海的主要航道。为了保证船只航行速度和河道畅通,特地环绕天恩桥西桥墩建此贴水平桥一座,民间称之为"桥里桥"或"一步登两桥"。这种设计,是几百年前的古代立交桥,在上海众多古石桥中绝无仅有,水乡江南也不多见。纤桥与天恩桥结合为一组集功能、技术、经济、美观于一体的优秀古石桥,体现了古代桥工的智慧和力量。

Qian Bridge, originally built in the Qing Dynasty and rebuilt during the Republican era, is a three-span beam bridge made of granite. It measures 18.7 meters in length and 1.6 meters in width, with a central span of 7 meters. The bridge deck is paved with rectangular flagstones, and the bridge piers are made of parallel rectangular flagstones, with stone bridge abutments at both ends. It is the only well-preserved stone slab bridge along the Hengli River in Jiading, which served as an important passage for boats and boatmen.

The Hengli River is connected to the south with the Wusong River and to the north with the Liuhe River, making it a busy waterway that has historically served as a major route from Jiading, Taicang, and other places to Shanghai. To ensure the smooth flow of the river and enable fast boat navigation, a special low-level bridge was constructed around the western pier of Tian'en Bridge, known as the "Bridge Inside Bridge" or "Two Bridges in One Step" by locals. This unique design is an ancient overpass bridge from several hundred years ago, rare not only in Shanghai but also in the water village of Jiangnan. Combined with Tian'en Bridge, Qian Bridge is an exceptional ancient stone bridge that embodies functionality, technology, economy, and aesthetics, reflecting the wisdom and strength of ancient bridge builders.

鹤槎山
Hecha Mountain

年代：宋至明
类别：古遗址
保护级别：嘉定区文物保护单位
利用情况：开放参观
Era: Song to Ming Dynasties
Category: Ancient ruins
Conservation level: Jiading district-level protected site
Utilization: Open for visit

鹤槎山，为南宋建炎三年（1129）浙江制置使韩世忠为阻击南犯金兵所筑的烽墩。当时韩世忠驻军吴淞江下游，设中军于江湾，为传递军情，沿走马塘一线构筑墩台营汛，六里至九里设一墩，东起江湾，西至嘉定，共18墩，以备瞭望，后成为江湾巡检司营汛驻地。因位于"白鹤南翔""槎溪逶流"之地，故后人取名为"鹤槎山"。

遗址现为圆形土堆，占地面积约1200平方米，黄土夯筑，上有民国时期建造的碉堡一座，南麓建有香雪庵，东麓有古银杏一棵。鹤槎山是上海地区现存极少的古代烽墩之一，对研究嘉定海陆变迁、古代军事设施、人文历史等具有重要价值。现嘉定尚存鹤槎山、方泰二处墩台遗址。

Hecha Mountain is a beacon tower that was built in the 3rd year of the Jianyan reign of the Southern Song Dynasty (1129) by Han Shizhong, the Zhejiang supervisor, to resist the invading Jin army from the north. During that time, Han Shizhong stationed his troops downstream of the Wusong River and established a central camp in Jiangwan. To transmit military information, he constructed beacon towers and camps along the Zouma Pond, with one tower every six to nine li. The line stretched from Jiangwan in the east to Jiading in the west, totaling 18 towers. Later on, it became the base for the Jiangwan Patrol Inspectorate. Due to its location in the area of "white cranes in Nanxiang" and "Chaxi flowing", it was named "Hecha Mountain" by later generations.

The site is now a circular earth mound covering an area of about 1200 square meters. It was built with rammed yellow soil and has a military fort constructed during the Republican period on top. To the south of the mountain's foot lies Xiangxue Temple, and to the east, an ancient ginkgo tree can be found. Hecha Mountain is one of the few remaining ancient beacon towers in the Shanghai area, holding historical value for studying changes in Jiaxing's land and sea, ancient military facilities, and human history. Currently, Jiaxing has two remaining beacon tower sites: Hecha Mountain and Fangtai.

练塘镇

Liantang Town

1 陈云故居
练塘镇下塘街 95 号

1 *Chen Yun Former Residence*
No.95, Xiatang Street, Liantang Town

2 颜安小学（老教室、杜衡伯纪念塔）
练塘镇下塘街 16 号

2 *Yan'an Elementary School (Old Classroom, Du Hengbo Memorial Tower)*
No. 16, Xiatang Street, Liantang Town

3 吴志喜故居
练塘镇东风街 93 弄 11 号

3 *Wu Zhixi Former Residence*
No.11, Lane 93, Dongfeng Street, Liantang Town

4 东区救火会
练塘镇东风街 11-12 号

4 *The East District Fire Brigade*
No.11-12, Dongfeng Street, Liantang Town

5 天光寺
练塘镇练东村泖口

5 *Tianguang Temple*
Maokou, Liantong Village, Liantang Town

6 阜康酱园
练塘镇前进街 138 号

6 *Fukang Sauce and Pickle Shop*
No.138, Qianjin Street, Liantang Town

7 练塘下塘街街廊
练塘镇下塘街 44 弄 19-33 号

7 *Jielang (Street Corridor) in Liantang Xiatang Street*
No.19-33, Lane 44, Xiatang Street, Liantang Town

8 顺德桥
练塘镇前进街

8 *Shunde Bridge*
Qianjin Street, Liantang Town

9 朝真桥
练塘镇前进街 46 号（北堍）

9 *Chaozhen Bridge*
No.46 (North end of the bridge), Qianjin Street, Liantang Town

图例 LEGENDS

全国重点文物保护单位
National priority protected site

上海市文物保护单位
Shanghai city-level protected site

区级文物保护单位
District-level protected site

区文物保护点
District-level protected place

其他景点
Other attractions

范围外景点
Out-of-range attractions

历史风貌区范围
Scope of historical district

游览路线
Touring route

朱枫连接线 Zhufeng Connected Road

北叶港 Beiye Port

天光寺
Tianguang Temple ⑤

老朱枫公路 Old Zhufeng Road

吴志喜故居
Wu Zhixi Former
Residence

③

东区救火会
The East District Fire Brigade

④

义学桥
Yixue Bridge

朝真桥
Chaozhen Bridge

永兴桥
Yongxing Bridge

②

算盘文化馆
Abacus Culture Hall

颜安小学（老教室、杜衡伯纪念塔）
Yan'an Elementary School
(Old Classroom, Du Hengbo
Memorial Tower)

皋康酱园
Fukang Sauce
and Pickle Shop

⑨

练塘艺术馆
Liantang Art Museum

⑥

① 陈云故居
Chen Yun Former Residence

⑦

练塘下塘街街廊
Jielang (Street Corridor)
in Liantang Xiatang Street

顺德桥
Shunde Bridge

⑧

陈云纪念馆
Chen Yun Memorial Hall

东泖江 Changyu River

上澜 Lanyu

东港塘 Dongtang Pond

钟联路 Zhonglian Road

东泖江 Dongjiang River

中心路 Zhonglian Center Road

老朱枫公路 Old Zhufeng Road

泖甸路 Maodian Road

练苇路 Lianwei Road

本图为位置示意，与实际尺寸不符
Illustration is not proportional to the actual scale

练塘镇
Liantang Town

交通指南：
地铁 17 号线至朱家角站，换乘公交青蒸线 / 青小线
至老朱枫公路练新路站下车，步行 400 米可达。

Transportation Guide:
Metro Line 17 to Zhujiajiao Station, transfer to
Bus Qingzheng Line/Qingxiao Line and get off at
Old Zhufeng Highway Lianxin Road Station, and
walk for 400 meters.

红色水乡，绿色生态

曲折蜿蜒的市河贯穿古镇东西，静静流淌，宁静祥和。市河两岸商铺民居林立，粉墙灰瓦，古朴幽静。练塘，褪去了市井繁华，悠然尘外，充满江南风情。

练塘镇旧名章练塘，位于今青浦区西南。相传孙权造战舰于青龙浦，于此张帆练兵，故为"张练"。又传五代闽国高州刺史章仔钧与妻练夫人居此，因而得名"章练"。自唐至明，练塘地区属苏州府吴江、长洲管辖。清初形成市廛，分属于元和、吴江、青浦三县合辖，其东市属元和县东吴下乡颜安里二十八都，故又旧称颜安里。乾隆年间因棉纺织业兴起而发展成大镇，民居稠密，百货具备。清末民初时练塘镇为谷米、棉纺织品集散地，上海及杭州、湖州、常熟来购米者极多，"每早市，乡人咸集，舟楫塞港，街道摩肩"。棉纺织业发达，由客商远销各地，"颇为西商所争购"。同时还制造和出售灌溉工具镶车，远近驰名。当时镇东西长九里，南北宽六里，分东、西两市，街巷纵横。宣统二年（1910），为便于管辖，将吴江、元和两县插花地归并青浦县，置章练塘区。1958年后，随青浦县划归上海市管辖。2001年年初，撤销原小蒸、蒸淀、练塘三镇建制，设立今练塘镇。

练塘为西郊重镇，地处交界，连两省三府五县，水路通畅，陆行不便，出行皆需摆渡，特殊的地理位置和交通，使之成为中国共产党早期革命活动和地下斗争的重要地区，先后涌现出陈云、高尔松、高尔柏、吴志喜等一批革命先辈，在此浴血斗争，为古镇练塘增添了红色光彩。

据地方志记载，旧时练塘有八景：三泖行帆、九峰列翠、塔院晓钟、天光古刹、明因夕照、圆通朝爽、西来揽秀、鹤荡渔歌，宜人的自然风光与人文胜景并存，史上颇受赞誉。至今练塘古韵犹存，古镇格局以市河为主轴，东西向延展。沿市河两侧，前进街、东风街和下塘街，依然保持着粉墙黛瓦、简小自然的江南水乡民居建筑风

格，"高屋窄巷对街楼，小桥流水人家处"，河道水系、街巷空间、建筑风貌整体保存较好。

练塘不但历史人文荟萃，史迹众多，而且非物质文化遗产丰富。现有区级以上非物质文化遗产共计6项，其中包括国家级非物质文化遗产1项，即吴歌(青浦田山歌)。上海市非物质文化遗址1项，即土布染织技艺。区级4项，分别为匍经、练塘糕团制作技艺、练塘茭白叶编结制作技艺、庆号习俗。同时，练塘以盛产茭白著称，素有"华东茭白第一镇"之誉，每年金秋，都会举办"茭白节暨古镇文化旅游节"，传承非遗文化，让人们品美食，悦美景，欣赏红色水乡、绿色生态、古色古镇。

不可移动文物资源

2010 年 7 月，练塘镇被公布为第五批中国历史文化名镇，其中历史文化风貌保护区面积 57.5 公顷，核心保护区面积 16.2 公顷。镇域内现有各级不可移动文物 35 处，包括上海市文物保护单位 2 处，区级文物保护单位 9 处，区文物保护点 24 处。练塘不但有类型多样的古建筑、近现代建筑，而且革命史迹丰富，是一座具有光荣革命传统的红色古镇。

A town of patriots, with green ecological landscape

The winding and meandering Shihe River runs through the ancient town from east to west, flowing calmly and peacefully. Along both sides of the river, there are numerous shops and residential buildings with powdered walls and grey tiles, presenting a quaint and tranquil atmosphere. Liantang, stripped of the hustle and bustle of the city, exudes a leisurely charm and embodies the essence of Jiangnan's allure.

Liantang Town, formerly known as Zhangliantang, is located in the southwestern part of present-day Qingpu District. According to legend, Sun Quan, the ruler of the Kingdom of Wu, constructed warships in Qinglongpu and trained soldiers here, hence the name "Zhanglian". It is also said that during the Five Dynasties period, Zhang Zijun, the Governor of Gaozhou in the Min Kingdom, and his wife, Madam Lian, lived here, thus giving it the name "Zhanglian". From the Tang Dynasty to the Ming Dynasty, the Liantang area was under the administration of Wujiang and Changzhou in Suzhou Prefecture. In the early Qing Dynasty, markets were established, jointly administered by Yuanhe, Wujiang, and Qingpu counties. The eastern market belonged to Yan'anli, a village in Dongwu Township of Yuanhe County, consisting of twenty-eight neighborhoods. Therefore, it was also known as "Yan'anli". During the Qianlong period of the Qing Dynasty, Liantang developed into a large town due to the rise of cotton textile industry, with densely populated residential areas and a wide variety of goods available. In the late Qing Dynasty and early Republic of China, Liantang became a distribution center for grain and cotton textiles, attracting many buyers from Shanghai, Hangzhou, Huzhou and Changshu. "Every morning, people from the countryside gathered, and the ports were crowded with boats and ships, making the streets bustling." The cotton textile industry flourished, and the merchants exported goods far and wide, making the town "a sought-after destination for Western merchants". Additionally, the town had a reputation for manufacturing and selling irrigation tools and carriages, earning a reputation locally and afar. During that time, the town covered a span of nine li from east to west and six li from north to south, and included two markets, East and West, and a network of streets and lanes. In the 2nd year of the Xuantong reign (1910), the Chahua District, which comprised

Wujiang and Yuanhe counties, was merged into Qingpu County and renamed Zhangliantang District for governance purposes. After 1958, it was placed under the jurisdiction of Shanghai municipality, as part of Qingpu County. In early 2001, the original Xiaozheng, Zhengdian, and Liantang towns were abolished and the current Liantang Town was established.

Liantang, located at the border, is a major town in the western part of the region, connecting two provinces, three prefectures, and five counties. It has convenient water transportation but inconvenient land transportation, requiring ferry services for travel. Due to its unique geographical location and transportation, Liantang was an important area for early revolutionary activities and underground struggles of the Communist Party of China. Many revolutionary predecessors such as Chen Yun, Gao Ersong, Gao Erbai, Wu Zhixi, etc., fought bloodily here, adding a revolutionary glory to the ancient town of Liantang.

According to local records, there were eight scenic spots in the old Liantang: the sailing boats on Sanmao River, the nine peaks adorned with emerald green, the dawn bell at the tower, the ancient monastery in the sunlight, the reflection of the setting sun in Mingyin Lake, the serene atmosphere of Yuantong Temple, the beauty of Xilai Gorge, and the fishing songs in Hedang Village. The pleasant natural scenery and cultural landscapes coexist and have received praise throughout history. The ancient charm of Liantang still lingers today, with the town's layout centered around the Shihe River, stretching from east to west. Along the banks of the river, Qianjin Street, Dongfeng Street, and Xiatang Street still preserve the architectural style of small, natural, traditional homes of the Jiangnan water villages, with their whitewashed walls and tiled roofs. The overall preservation of the river system, street and alley spaces, and architectural style is relatively good.

Liantang not only has a rich history and cultural heritage but also possesses a wealth of intangible cultural heritage. Currently, there are a total of 6 intangible cultural heritage items at the district level or above. Among them, there

练塘镇参观指南 Liantang Town Visiting Guide

游览路线：
陈云纪念馆→陈云故居→算盘文化馆→练塘艺术馆→永兴桥→颜安小学（老教室、杜衡伯纪念塔）→义学桥→东区救火会→朝真桥→阜康酱园

Tourist Route:
Chen Yun Memorial Hall → Chen Yun Former Residence → Abacus Culture Hall → Liantang Art Museum → Yongxing Bridge → Yan'an Elementary School (Old Classroom, Du Hengbo Memorial Tower) → Yixue Bridge → The East District Fire Brigade → Chaozhen Bridge → Fukang Sauce and Pickle Shop

古镇美食特产：
糕团、茭白、土布

Local Specialties:
Gaotuan, Jiao Bai, Homespun (handwoven cloth)

is a national-level intangible cultural heritage item, which is Wu Ge (Farm Song of Qingpu). Additionally, there is a municipal level item, which is the technique of mud dyeing and weaving. Furthermore, there are 4 district-level items, including Pu Jing (a type of massage technique used in traditional folk medicine), the production technique of Liantang Gaotuan (a kind of traditional cake), the production technique of the leaves of Zizania latifolia (a kind of leaf weaving) in Liantang, and the tradition of Qing Hao (celebratory calligraphy marking auspicious occasions). Moreover, Liantang is well-known for its abundant production of Jiao Bai (a type of edible bamboo shoot), earning it the title of being the "number one town for Jiao Bai in East China". Every year in the golden autumn, the "Jiao Bai Festival and Ancient Town Cultural Tourism Festival" takes place, showcasing and preserving intangible cultural heritage, offering delicious food, and allowing visitors to appreciate the town's the red water village, green ecology, and ancient charm of the town.

Immovable Cultural Relics Resources

In July 2010, Liantang Town was announced as the fifth batch of national famous historical and cultural towns, with a historical and cultural conservation area covering 57.5 hectares, and a core protection area spanning 16.2 hectares. Within the town, there are currently 35 immovable cultural relics at various levels, including 2 Shanghai city-level protected sites, 9 district-level protected sites and 24 district-level protected places. Liantang not only boasts diverse types of historic buildings and modern architecture, but also has a rich revolutionary history, making it a historic red town with a glorious revolutionary tradition.

陈云故居
Chen Yun Former Residence

年代：清代
类别：近现代重要史迹及代表性建筑
保护级别：上海市文物保护单位
利用情况：开放参观

Era: Qing Dynasty
Category: Modern important historical sites and representative buildings
Conservation level: Shanghai city-level protected site
Utilization: Open for visit

陈云同志少年时期的住所，原为陈云舅父廖文光的家宅。故居分一层店面和二层居室两个部分，中间为小天井。店面北面临街，是当时廖文光夫妇开设裁缝铺和小酒肆之处；店面后即居室，楼上为陈云舅父母所居，楼下为陈云居住的房间。

故居坐南朝北，砖木结构，一开间，面阔 4.35 米，进深 18.5 米，总建筑面积 96 平方米。硬山顶，上铺小青瓦。店面为七架梁，穿斗式梁架，北为店门，南有槛窗。居室为七架梁，穿斗式梁架。故居内的陈设基本复原了当年的面貌。

Chen Yun Former Residence is the childhood home of comrade Chen Yun. At first, it was the home of Chen Yun's uncle, Liao Wenguang. The former residence comprises a ground-floor storefront and a second-floor living area, with a small courtyard in between. The storefront is facing the north side of the street and was where Liao Wenguang and his wife operated a tailor shop and a small tavern. The living quarters were situated behind the storefront, with Chen Yun residing on the ground floor while his uncle and aunt living upstairs.

The former residence faces north direction and has a brick and wood structure. It consists of a single room with dimensions of a width of 4.35 meters and a depth of 18.5 meters, totaling a building area of 96 square meters. The roof is a flush gable roof covered with small gray tiles. The storefront features 7-purlin beams with column and tie construction, with the entrance to the shop on the north side and threshold windows on the south side. The living area also has 7-purlin beams with column and tie construction. Overall, the interior of the former residence has largely retained its original appearance.

颜安小学（老教室、杜衡伯纪念塔）

Yan'an Elementary School (Old Classroom, Du Hengbo Memorial Tower)

年代： 清代（颜安小学老教室），近现代（杜衡伯纪念塔）
类别： 古建筑，近现代重要史迹及代表性建筑
保护级别： 青浦区文物保护单位
利用情况： 教育场所

Era: Qing Dynasty (Yan'an Elementary School Old Classroom); Modern Times (Du Hengbo Memorial Tower)
Category: Ancient architecture; Modern important historical sites and representative buildings
Conservation level: Qingpu district-level protected site
Utilization: Educational site

颜安小学，前身是颜安书院，创建于清光绪十五年（1889），由镇人捐资利用原同仁堂建筑改建而成，因地处练塘古镇东部的颜安里而得名，取"颜子安贫乐道"之意。中国共产党人陈云、吴志喜、高尔松、高尔柏等皆毕业于该校。

校内现存陈云等少年时就读的老教室一幢和杜衡伯纪念塔一座。老教室位于学校南侧，坐南朝北，砖木结构，七架梁，六开间，通面阔 18.75 米，通进深 6.15 米，建筑面积 115 平方米。杜衡伯纪念塔位于教室西北面、通廊南侧，始建于 1936 年，由陈云、高尔松等 40 人集资建造，为方形基座，长 0.65 米，高 0.5 米，塔身高 1.6 米，梯台式，正面题"杜校长衡伯先生纪念塔"。

杜衡伯（1891—1934），又名杜枢，青浦人，进步教育家，公立颜安小学的首任校长，任职 16 年，办学成绩卓著。杜衡伯安排陈云同志免费进入颜安小学读高小部至毕业，是陈云的恩师。

Yan'an Elementary School, formerly known as Yan'an Academy, was established in the 15th year of the Guangxu reign of the Qing Dynasty (1889). It was renovated from the original Tongrentang building with funds donated by the local residents. The school is named after its location in the eastern part of Liantang Ancient Town, known as Yan'anli, and embodies the spirit of "Yan Zi promotes the way with humility and joy despite his poverty". Prominent members of the Communist Party of China, such as Chen Yun, Wu Zhixi, Gao Ersong, and Gao Erbai, all graduated from this school.

The old classroom where Chen Yun and others studied and the Du Hengbo Memorial Tower, still exist on the school's premises. Located on the south side of the school and facing north, the old classroom is a brick and wood structure with 7-purlin beams and six compartments. It has a width of 18.75 meters and a depth of 6.15 meters, with a total building area of 115 square meters. On the south side of the corridor, northwest of the classroom, stands the Du Hengbo Memorial Tower. Constructed in 1936, the funds for its construction were contributed by Chen Yun, Gao Ersong, and 40 others. The tower is square-shaped with a base length of 0.65 meters and a height of 0.5 meters. It stands at a height of 1.6 meters and features a ladder platform. On the front of the tower, the inscription reads "Du Principal Hengbo Memorial Tower".

Du Hengbo (1891–1934), also known as Du Shu, was a progressive educator from Qingpu. He served as the first principal of the public Yan'an Elementary School for 16 years and achieved remarkable results in education. Du Hengbo arranged for Comrade Chen Yun to study at Yan'an Elementary School's upper-primary section free of charge until graduation, and he was Chen Yun's mentor.

吴志喜故居
Wu Zhixi Former Residence

年代：清代
类别：近现代重要史迹及代表性建筑
保护级别：青浦区文物保护点
利用情况：居住场所
Era: Qing Dynasty
Category: Modern important historical sites and representative buildings
Conservation level: Qingpu district-level protected place
Utilization: Residence

　　吴志喜（1911—1928），练塘镇人，革命烈士。曾在青浦、枫泾参与和组织策划革命活动，1928 年牺牲于松江小校场。故居建于清代，二层砖木结构，坐北朝南，二开间，面阔 8.12 米，进深 8.35 米，九架梁，硬山顶，上铺小青瓦。一层南、北面每间原有 6 扇长窗，现已不存。二层南、北面每间为 8 扇槛窗，有破损。

Wu Zhixi (1911–1928) was a revolutionary martyr from Liantang Town. He participated in and organized revolutionary activities in Qingpu and Fengjing, and died a heroic death in Songjiang small drill ground in 1928. Wu Zhixi Former Residence was built during the Qing Dynasty. It is a two-story brick and wood structure, facing south, with two bays. The width is 8.12 meters and the depth is 8.35 meters. It has 9-purlin beams, a flush gable roof covered with small gray tiles. On the first floor, there were originally six long windows in each room on the south and north sides, but they no longer exist. On the second floor, each room on the south and north sides has eight lattice windows, some of which are damaged.

东区救火会
The East District Fire Brigade

年代： 清代
类别： 近现代重要史迹及代表性建筑
保护级别： 青浦区文物保护点
利用情况： 居住场所
Era: Qing Dynasty
Category: Modern important historical sites and representative buildings
Conservation level: Qingpu district-level protected place
Utilization: Residence

民国时期利用惠世庵旧址设置，承担练塘镇东片的消防工作。惠世庵是明嘉靖四十二年（1563）里人万麟捐宅创建，清咸丰三年（1853）辟为惠世义学，光绪年间修山门及正殿。该建筑坐北朝南，砖木结构，前后二进。前进面阔二间，临街，东侧门额上写有"东区救火会"五字，西侧门额上写有"青浦县公安第二分局"九字。后进面阔二间，七架梁，中间有一天井，硬山顶，小瓦屋面，雌毛脊。

The East District Fire Brigade was established in the former site of Huishi Temple during the Republican era and was responsible for firefighting in the eastern part of Liantang Town. Huishi An was originally built in the 42nd year of the Jiajing reign of the Ming Dynasty (1563) by Wan Lin, a native. In the 3rd year of the Xianfeng reign of the Qing Dynasty (1853), it was converted into Huishi Yixue(a traditional school in Chinese education system), and the gate and main hall were reconstructed during the Guangxu reign. The building faces south and has a brick and wood structure with two sections in the front and back. The front section has a width of two bays and faces the street. On the east side of the entrance, there is a plaque with the words "East District Fire Brigade", and on the west side, there is a plaque with the words "Second Branch of Qingpu County Public Security Bureau". The back section has a width of two bays, 7-purlin beams, and a courtyard in the middle. It has a flush gable roof covered with small tiles, and a ridge adorned with Cimao ridge (a type of decoration on the ridge end of traditional Chinese architecture, it is usually the form of an eagle's bea which is made of gray plastic).

天光寺
Tianguang Temple

年代：宋代
类别：古建筑
保护级别：青浦区文物保护单位
利用情况：宗教活动
Era: Song Dynasty
Category: Ancient architecture
Conservation level: Qingpu district-level protected site
Utilization: Religious activities

相传五代章仔钧与妻章练夫人居于练塘，后举家赴福建赴任，遂舍宅为寺，即为"天光寺"。宋端平年间（1234—1236）重建，明弘治丁巳及天启甲子两次修葺，至清道光二十七年（1847）又重修。现存天王殿和大雄宝殿，均为三开间，九架梁，歇山顶。其中天王殿面阔11.3米，进深8.9米，正立面次间有"双龙戏珠"漏窗。大雄宝殿面阔13.3米，进深11.8米。

According to legend, Zhang Zijun and his wife, Lady Zhang Lian, lived in Liantang during the Five Dynasties period. They later moved to Fujian for official duties, and their former residence was converted into a temple known as "Tianguang Temple". The temple was reconstructed during the Duanping period of the Song Dynasty (1234–1236) and underwent renovations twice during the reigns of Hongzhi in the Ming Dynasty and Tianqi in the Qing Dynasty. The temple was further repaired in the 27th year of the Daoguang reign (1847). Currently, the temple retains the Tianwang Hall and the Mahavira Hall. Both halls consist of three bays, 9-purlin beams, and gable and hip roof. The Tianwang Hall measures 11.3 meters wide and 8.9 meters deep, featuring "twin dragons playing with a pearl" carved leaking window on its front façade. The Mahavira Hall measures 13.3 meters wide and 11.8 meters deep.

阜康酱园
Fukang Sauce and Pickle Shop

年代：近现代
类别：近现代重要史迹及代表性建筑
保护级别：青浦区文物保护点
利用情况：居住场所
Era: Modern times
Category: Modern important historical sites and
representative buildings
Conservation level: Qingpu district-level protected place
Utilization: Residence

　　阜康酱园，为庭院式建筑，创办于1919年，所产的酒、酱、饴糖、酱菜远近闻名。酱园坐北朝南，砖木结构。青砖外墙，墙上写有红色"酱园"两个大字，墙檐有砖饰，设木板门，门上镶有铜泡钉353枚。正屋两层，三开间，九架梁，硬山顶，檐口有木雕饰，小瓦屋面。

　　Fukang Sauce and Pickle Shop is a courtyard-style building founded in 1919. It is famous for its production of liquor, soy sauce, yitang (a type of candy), and preserved vegetables. The building faces north and is built with brick and wood structure. The exterior walls are made of green bricks with the words "Sauce and Pickle Shop" written in red. The eaves of the walls are adorned with brick decorations, and the wooden doors have 353 copper nail studs. The main building has two floors, with three bays and 9-purlin beams. It features a gable and hip roof. The eaves are adorned with wooden carvings, and the roof is covered with small tiles.

练塘下塘街街廊

Jielang (Street Corridor) in Liantang Xiatang Street

年代：近现代
类别：近现代重要史迹及代表性建筑
保护级别：青浦区文物保护点
利用情况：居住场所
Era: Modern times
Category: Modern important historical sites and
representative buildings
Conservation level: Qingpu district-level protected place
Utilization: Residence

　　街廊建于民国时期，沿李华港构筑，坐西朝东，砖木结构，西面段已缺失，现存 14 间，通面阔 43.9 米，进深 12.2 米，两层，重檐，东面原有廊柱 14 根，后增加 14 根，以加固支撑。硬山顶，哺龙脊上铺小青瓦，南侧有五山封火墙。

　　The Jielang, or street corridor, in Liantang Xiatang Street, was built during the Republican era, along the Lihua Port. It faces east and is built with brick and wood structure. The western section is missing, but there are currently 14 sections remaining. The corridor has a width of 43.9 meters and a depth of 12.2 meters. It consists of two floors with a double eave design. Originally, there were 14 columns on the east side, but 14 additional columns were later added for increased support. It has flush gable roof, with dragon-shaped ridge covered by small gray tiles, and there is a Wushan (five mountains) firewall on the south side.

顺德桥
Shunde Bridge

年代：元代
类别：古建筑
保护级别：青浦区文物保护单位
利用情况：开放参观
Era: Yuan Dynasty
Category: Ancient architecture
Conservation level: Qingpu district-level protected site
Utilization: Open for visit

元至正三年（1343）始建，清顺治年间（1644—1661）、康熙五十八年（1719）、乾隆四十九年（1784）多次重修，三跨平梁桥，南北向，跨市河三里塘。桥长 17.5 米，宽 2.4 米。桥柱由 3 根石柱并立而成，上置横梁，再置 6 根楠木，桥前端用长条石铺搁开楠木上，两岸桥墩亦用条石并立，两侧有护栏、望柱及抱鼓石，具有元代梁桥典型特征。

Shunde Bridge was initially built in the 3rd year of the Yuan Dynasty's Zhizheng reign (1343). It underwent several renovations during the Qing Dynasty, specifically during the Shunzhi reign (1644–1661), the 58th year of the Kangxi reign (1719), and the 49th year of the Qianlong reign (1784). It is a three-span flat beam bridge, oriented north-south, spanning the Sanlitang section of the Shihe River. The bridge is 17.5 meters long and 2.4 meters wide. The bridge columns are formed by three standing stone pillars, with a horizontal beam placed on top. Six nanmu beams are then placed on top of the horizontal beam. At the front end of the bridge, a long stone strip is laid on top of the nanmu beams. The bridge piers on both sides are also made of stacked stone blocks. The bridge is adorned with railings, wangzhu-balustrade pillars, and drum-shaped stones on both sides of the bridge deck, showcasing typical characteristics of Yuan Dynasty bridges.

朝真桥
Chaozhen Bridge

年代: 明代
类别: 古建筑
保护级别: 青浦区文物保护单位
利用情况: 开放参观
Era: Ming Dynasty
Category: Ancient architecture
Conservation level: Qingpu district-level protected site
Utilization: Open for visit

始建年代不详，明嘉靖三十四年（1555）里人富商万麟重建，清康熙三十四年（1695）又重建，因北侧原有朝真道院（俗名圣堂）而俗称"圣堂桥"。该桥为单孔石拱桥，南北向，跨市河，青石、花岗石材质。桥长 18 米，宽 3.45 米，跨径 6.82 米，桥面有镂空青石护栏。两侧桥柱上镌有楹联，东联为"帝泽流通万里，神庥绵亘千秋"，西联为"长留题柱高联，重记受书胜事"。

The Chaozhen Bridge's exact construction date is unknown, but it was rebuilt by a wealthy local merchant named Wan Lin in the 34th year of the Ming Dynasty's Jiajing reign (1555). It was rebuilt again in the 34th year of the Qing Dynasty's Kangxi reign (1695). The bridge is commonly known as "Shengtang Bridge" because there was a Chaozhen Daoist Temple (commonly known as Shengtang) on the northern side. It is a single-arch stone bridge, oriented north-south, spanning the Shihe River. It is constructed with blue stone and granite. The bridge is 18 meters long, 3.45 meters wide, with a span of 6.82 meters. The bridge has a hollow blue stone railing. There are couplets carved on the bridge pillars on both sides. The eastern couplet reads, "The emperor's blessings flow for thousands of li, and divine spirits extend through the ages". The western couplet reads, "Stay long and inscribe the high couplets, recounting the triumphs of receiving sacred texts".

张堰镇

Zhangyan Town

1 **姚光故居**
张堰镇新建路 130 号

2 **钱家祠堂**
张堰镇东河沿路 15 号

3 **卢家祠堂**
张堰镇新建路 65 号

4 **陈氏宅**
张堰镇政安弄 26 号

5 **钱培名宅**
张堰镇政安弄 3 号

6 **白蕉故居**
张堰镇新尚路 16 号；东大街 134 弄 2-4 号

7 **吴梁三命坊**
张堰镇张堰大街 441 号

1 *Yao Guang Former Residence*
No.130, Xinjian Road, Zhangyan Town

2 *Qian Family Ancestral Hall*
No.15, Dongheyan Road, Zhangyan Town

3 *Lu Family Ancestral Hall*
No.65, Xinjian Road, Zhangyan Town

4 *Chen Family Residence*
No.26, Zheng'an Lane, Zhangyan Town

5 *Qian Peiming Residence*
No.3, Zheng'an Lane, Zhangyan Town

6 *Bai Jiao Former Residence*
No.16 Xin Shang Road; No. 2-4, Lane 134, East Street, zhangyan Town

7 *Wu Liang San Ming Memorial Archway*
No.441, Zhangyan Street, Zhangyan Town

图例 LEGENDS

- 全国重点文物保护单位
  National priority protected site
- 上海市文物保护单位
  Shanghai city-level protected site
- 区级文物保护单位
  District-level protected site
- 区文物保护点
  District-level protected place
- 其他景点
  Other attractions
- 历史风貌区范围
  Scope of historical disctrict
- 游览路线
  Touring route

石水路 Shishui Road

东门路 Dongmen Road

留溪路 Liuxi Road

白蕉故居
Bai Jiao
Former Residence

张堰历史人文风情馆
Zhangyan History and Humanities Style Museum

张堰公园
Zhangyan Park

花贤路 Huaxian Road

新华路 Xinhua Road

牛桥港 Niuqiao Port

新贝街 Xinbei Street

白蕉弄 Shijie Lane

陈氏宅
Chen Family Residence

白蕉故居
Bai Jiao Former Residence

钱家祠堂
Qian Family
Ancestral Hall

白蕉艺术馆
Bai Jiao Art Museum

姚光故居
Yao Guang Former Residence

新港河 Xingang River

钱培名宅
Qian Peiming Residence

吴梁三命坊
Wu Liang San Ming Memorial Archway

卢家祠堂
Lu Family Ancestral Hall

松金公路 Songjin Highway

张泾河 Zhangjing River

本图为位置示意，与实际尺寸不符
Illustration is not proportional to the actual scale

张堰镇

Zhangyan Town

交通指南:
地铁 1 号线至莲花路站，换乘公交莲卫专线至东贤路松金公路站，步行 10 分钟可达；
地铁 10 号线至虹桥火车站，换乘公交虹桥枢纽 7 路至松金公路张堰站，步行 15 分钟可达。

Transportation Guide:
Metro Line 1 to Lianhua Road Station, transfer to Bus Lianwei Special Line to Dongxian Road Songjin Highway Station, 10 minutes' walk; Metro Line 10 to Hongqiao Railway Station, transfer to Bus Hongqiao Junction No.7 to Songjin Highway Zhangyan Station, 15 minutes' walk.

浦南首镇，文教流长

张堰镇位于金山区东南部，古称赤松里，相传汉留侯张良从赤松子游曾居于此地，故又称留溪、张溪。历史上的张堰因其优越的地理位置、发达的交通和繁华的商业，被后人誉为"浦南首镇"。同时，它也是一处文教宝地和革命高地，明清两代有 23 位进士籍居张堰，近代有全国性革命文学团体——"南社"在此孕育，诺贝尔物理学奖得主、"光纤之父"高锟，著名书画家白蕉等名人大家也出自张堰。其传承已久的文教传统和敢于斗争的革命基因赋予了张堰深刻的历史文化内涵，使张堰成为了兼具传统人文情怀与江南地域特色的文化名镇。

张堰地区在春秋时期已有村落，唐末五代为御海潮置华亭十八堰，其中之一为张泾堰，镇袭堰名，俗称张堰，时设浦东场大使署于镇。明代，张堰以南设金山卫，"商贾畏军强，莫敢往卫"，张堰一度获得长足发展，"镇遂盛"。镇上"南湖头"曾是重要的商业地标，

商船聚泊，樯桅林立，烟火之盛，甲于一方。商业繁华，各类商店、作铺鳞次栉比，是金山、平湖、奉贤一带商业汇集之地。清代，镇设有金山分府署，镇上工商业发达。到清末民初时期，金山县内最早的钱庄、银行、商会相继在张堰镇建立。同时，清末的张堰还成为金山县革命的策源地。中国共产党成立后，恽代英、萧楚女、陈云等革命家都曾来镇指导革命活动。1924 年国共合作时期的国民党金山县党部始建于镇，金山县第一位中国共产党员李一谔及一批革命人士于 1926 年在张堰成立中共浦南特支，张堰镇成为中国共产党早期在金山南部地区的活动中心。

历史上的张堰曾经水网纵横，是典型的源于自然的紧密城镇，有大小弄巷 29 条，建筑风貌紧凑，呈现出许多小而精致的院落空间和公共空间，曾有"板桥夜眺""秦峰一翠""留溪观潮"等"张堰八景"。数百年后的今天虽然河道有所填埋，建筑几经换代，但保留有较为完整的街巷空间肌理。镇内现有四片保存较好的明清

建筑群：南社纪念馆建筑群、石皮弄建筑群、政安弄建筑群、西河沿建筑群。这些建筑以一、二层为主，风格较为统一，依稀可见昔日的繁荣景象。另有梁坊、古桥等具有江南传统环境特色的典型建筑点缀其中，构成了张堰镇传统江南水乡的历史风貌。

此外，张堰还保存有金山堰菜、张堰李氏骨伤科医疗技法、"闻万泰"酱菜制作技艺、张堰喜庆点心制作技艺等非物质文化遗产，这些传统技艺植根张堰的乡土文化中，反映了张堰人代际相传的民间习俗和传统技艺。

如今的张堰通过活化历史建筑，打造了上海南社纪念馆、白蕉艺术馆、上海市华侨书画院、大隐书局等一批兼具传统人文艺术与江南地域特色的文化体验场所，让游客到张堰有古迹可游、有非遗可玩、有美食可品，在娱乐休憩中体验传统文化的魅力。

不可移动文物资源

2010 年 7 月，张堰镇被公布为第五批中国历史文化名镇，其中历史文化风貌保护区面积 41.39 公顷，核心保护区面积 11.8 公顷。镇域内现有各级不可移动文物 20 处，包括上海市文物保护单位 1 处，区级文物保护单位 5 处，区文物保护点 14 处。建筑年代主要集中在清末与民国时期，类型丰富，含民居、古桥、梁坊、碑刻等。

张堰镇参观指南 Zhangyan Town Visiting Guide

游览路线：
张堰历史人文风情馆→张堰公园→白蕉艺术馆→陈氏宅（陈家走马楼）→钱培名宅（朱鹏高艺术馆）→卢家祠堂（大境堂）→姚光故居（上海南社纪念馆）

Tourist Route:
Zhangyan History and Humanities Style Museum → Zhangyan Park → Bai Jiao Art Museum → Chen Family Residence (Zou Ma Lou) → Qian Peiming Residence (Zhu Penggao Art Museum) → Lu Family Ancestral Hall (Dajing Hall) → Yao Guang Former Residence (Shanghai Nan She Memorial Hall)

古镇美食特产：
堰菜、"闻万泰"酱菜

Local Specialties:
Yancai, "Wen Wan Tai" Pickled Vegetables

A town prominent in Punan, with a long history of education and culture

Zhangyan Town is located in the southeastern part of Jinshan District, formerly known as Chisongli. According to legend, Zhang Liang, Marquis Liuhou (Marquis of Pingyang) from the Han Dynasty, once resided in this area, which is why it is also called Liuxi or Zhangxi. Throughout history, Zhangyan has earned the reputation of being the "foremost town in Punan" due to its advantageous geographical location, developed transportation, and thriving commercial activities. At the same time, it is also a place of cultural and educational significance as well as a revolutionary stronghold. During the Ming and Qing dynasties, 23 successful candidates of the imperial examination were from Zhangyan. In modern times, the renowned national revolutionary literary group "Nan She" was formed here. Famous individuals such as Nobel laureate and "Father of Fiber Optics" Charles K. Kao, and renowned painter and calligrapher Bai Jiao also hail from Zhangyan. Its long-standing tradition of culture and education, as well as its revolutionary spirit, have endowed Zhangyan with profound historical and cultural connotation making it a cultural famous town that combines traditional humanistic feelings with the unique characteristics of the Jiangnan region.

During the Spring and Autumn period, there were already villages in the Zhangyan area. In the late Tang Dynasty and the Five Dynasties, the Imperial Court established eighteen dams and canals for controlling the ocean tide, one of which was Zhangjing Canal. The town took on the name of the canal and became commonly known as Zhangyan. During that time, the Pudong "Ambassador Office" (a local administrative institution) was established in the town. In the Ming Dynasty, the Jinshan Wei was established to the south of Zhangyan. "The merchants feared the strong military presence", which brought about a rapid development of Zhangyan, "becoming a flourishing town". The "Nanhutou" in the town was an important commercial landmark where merchant ships gathered and the streets were lined with masts and sails. It was a vibrant scene of economic prosperity and became a commercial center for Jinshan, Pinghu, and Fengxian. In the Qing Dynasty, the town had a separate Jinshan Prefecture Office and its industry and commerce flourished. By the late Qing Dynasty and early Republic of China period, the earliest money house, bank, and business association in Jinshan County were established in Zhangyan. At the same time, Zhangyan in the late Qing Dynasty became a source of revolution in Jinshan County. After the establishment of the Communist Party of China, revolutionaries such as Yun Daiying, Xiao Chunu, and Chen Yun came to the town to guide revolutionary activities. The Kuomintang's Jinshan County Party Committee was established in the town during the period of Kuomintang-Communist cooperation in 1924. The first Communist Party of China member in Jinshan County, Li Yi'e, and a group of revolutionaries established the CPC Punan Special Branch in Zhangyan in 1926, making Zhangyan Town the activity center of the early Communist Party of China in the southern part of Jinshan.

In history, Zhangyan was a town characterized by a network of waterways, typical of a closely-knit settlement that originated from nature. There were 29 small and large alleyways, with compact architectural styles. It boasted numerous small and exquisite courtyard spaces and public spaces. There were also famous "Eight Scenic Spots of Zhangyan" such as "Night View of Ban Bridge" "Greenery of Qinfeng Mountain" and "Tidal Watching at Liuxi", which showcased the beauty and charm of the town. Even though hundreds of years have passed, today Zhangyan still retains a relatively intact street and alley spatial structure, although the waterways have been partially filled and the buildings have undergone several renovations. There are four well-preserved architectural groups from the Ming and Qing dynasties in the town: the buildings of the Nan She Memorial Hall, the Shipi Alley, the Zheng'an Alley, and the Xihe Yan. These buildings mainly consist of one or two floors and have a relatively unified style, faintly reflecting the past prosperity. In addition, there are typical architectural features of the traditional Jiangnan environment, such as Liang Fang and ancient bridges, which adorn the town and constitute the historical charm of Zhangyan as a traditional Jiangnan water town.

In addition, Zhangyan also preserves intangible cultural heritage such as Jinshan Yancai (a local cuisine), the orthopedic medical techniques of the Li family in Zhangyan, the production techniques of "Wen Wan Tai" pickled vegetables, and the production techniques of festive pastries in Zhangyan. These traditional skills are rooted in the local culture of Zhangyan and reflect the intergenerational folk customs and traditional techniques passed down by the people of Zhangyan.

Today, Zhangyan has revitalized its historical buildings and created a series of cultural experience sites that combine traditional humanistic characteristics with the unique features of the Jiangnan region. These include the Shanghai Nan She Memorial Hall, the Bai Jiao Art Museum, the Shanghai Overseas Chinese Painting and Calligraphy Academy, and the Dayin Bookmall. They offer visitors the opportunity to explore historical sites, engage in intangible cultural heritage activities, and savor local delicacies, allowing them to experience the charm of traditional culture while enjoying leisure and entertainment.

Immovable Cultural Relics Resources

In July 2010, Zhangyan Town was announced as the fifth batch of national famous historical and cultural towns. The historical and cultural landscape protection area covers an area of 41.39 hectares, with a core protection area of 11.8 hectares. There are currently 20 immovable cultural relics at all levels in the town, including 1 Shanghai city-level protected sites, 5 district-level protected sites and 14 district-level protected places. The buildings mainly date back to the late Qing Dynasty and the Republican period, with a diverse range of types including residential buildings, ancient bridges, archways, and stone inscriptions.

姚光故居
Yao Guang Former Residence

年代：清代
类别：近现代重要史迹及代表性建筑
级别：上海市文物保护单位
利用情况：开放参观
Era: Qing Dynasty
Category: Modern important historical sites and representative buildings
Conservation level: Shanghai city-level protected site
Utilization: Open for visit

　　姚光，字凤石，号石子、复庐等，曾担任南社主任。故居系姚家祖宅，始建于明清，清咸丰年间（1851—1861）毁于兵灾，后姚光先祖重建，具体年代不详。故居为硬山顶，坐北朝南，黑瓦白墙，是典型的清末民初江南民居院落建筑风格。内设厅堂、天井、穿堂、厢房、后院等。姚光故居是姚光的居所，也是南社人士联络及会晤交流的一个重要场所，现开辟为上海南社纪念馆。

Yao Guang, with the courtesy name Fengshi, had other pen names such as Shizi and Fulu. He was a prominent figure who served as the director of the Nan She. His former residence, known as Yao Guang Former Residence, was the ancestral home of the Yao family and was initially built during the Ming and Qing dynasties. However, it was destroyed during a military disaster in the Xianfeng period of the Qing Dynasty (1851–1861). It was later rebuilt by Yao Guang's ancestors, but the exact date of reconstruction is unknown. Yao Guang Former Residence is a traditional Jiangnan-style courtyard facing south, having white walls and flush gable roof covered with black tiles. It faces south. The residence consists of halls, courtyards, connecting halls, wing rooms, and a backyard. This architectural style is typical of the late Qing Dynasty and early Republic of China period. Yao Guang Former Residence served as Yao Guang's dwelling and also played a vital role as a place for communication and meetings among Nan She members. It has been converted into the Shanghai Nan She Memorial Hall.

钱家祠堂
Qian Family Ancestral Hall

年代: 清代
类别: 古建筑
级别: 金山区文物保护单位
利用情况: 开放参观
Era: Qing Dynasty
Category: Ancient architecture
Conservation level: Jinshan district-level protected site
Utilization: Open for visit

钱家祠堂坐东北朝西南，砖木结构，四合院式布局，现存后厅及两侧厢房，共 17 间，建筑面积约 352 平方米。该祠堂所属张堰钱家，系钱圩钱氏一脉。钱家为金山望族，是清代有名的校勘世家。

The Qian Family Ancestral Hall faces southwest, and is a brick and wood structure. It has a courtyard-style layout, with the surviving rear hall and two side wing rooms, totaling 17 rooms, with a building area of approximately 352 square meters. The ancestral hall belongs to the Qian family from Zhangyan, who are descendants of the Qian clan in Qianwei. The Qian family is a prominent family in Jinshan and was a well-known family of textual critics in the Qing Dynasty.

卢家祠堂
Lu Family Ancestral Hall

年代：清代
类别：古建筑
级别：金山区文物保护点
利用情况：开放参观
Era: Qing Dynasty
Category: Ancient architecture
Conservation level: Jinshan district-level protected place
Utilization: Open for visit

　　卢家祠堂建于清代后期，是大体量、大进深、单层高敞的独栋祠堂，在上海较为少见。原主人卢道昌在清代曾任户部山东司主事，咸丰年间参与督办江苏团练。1937年8月，由方冲之、任道远等人发起，借卢家祠堂创办张堰初中补习班，招收学生五十余人。该补习班即为张堰中学的前身。卢家祠堂现为上海市华侨书画院张堰创作基地"大境堂"。

The Lu Family Ancestral Hall was built in the late Qing Dynasty and is a large, deep, single-story ancestral hall, which is relatively rare in Shanghai. The original owner, Lu Daochang, served as an official in the Shandong Branch of the Ministry of Household Affairs in the Qing Dynasty and was involved in supervising the Jiangsu Training Corps during the Xianfeng period. In August 1937, initiated by Fang Chongzhi, Ren Daoyuan, and others, the Lu Family Ancestral Hall was used to establish a supplementary class for Zhangyan Junior High School, enrolling more than fifty students. This supplementary class was the precursor of Zhangyan Middle School. The Lu Family Ancestral Hall is now the Shanghai Overseas Chinese Painting and Calligraphy Academy Zhangyan Creation Base known as "Dajing Hall".

陈氏宅
Chen Family Residence

年代： 清代
类别： 古建筑
级别： 金山区文物保护点
利用情况： 开放参观
Era: Qing Dynasty
Category: Ancient architecture
Conservation level: Jinshan district-level protected place
Utilization: Open for visit

　　陈氏宅始建于晚清时期，俗称陈家走马楼。前后二进二层楼房，天井居中，楼屋呈"回"字结构，楼上四周有走廊可贯通，是张堰为数不多的走马楼式建筑。2012 年镇政府出资重新修缮了陈氏宅，现开辟为"金邑传集——文创雅集生活空间"。

　　The Chen Family Residence was originally built in the late Qing Dynasty and is commonly known as the Chen Family "Zou Ma Lou". It consists of two sections and two floors, with a courtyard in the center. The building is structured in the shape of the Chinese character " 回 " (hui), with corridors connecting all four sides on the upper floor. It is one of the few Zou Ma Lou style buildings in Zhangyan. In 2012, the local government funded the restoration of the Chen Family Residence, which is now opened as the "Jin Yi Chuan Ji—Cultural and Creative Elegant Living Space".

钱培名宅
Qian Peiming Residence

年代： 清代
类别： 近现代重要史迹及代表性建筑
级别： 金山区文物保护点
利用情况： 开放参观
Era: Qing Dynasty
Category: Modern important historical sites and representative buildings
Conservation level: Jinshan district-level protected place
Utilization: Open for visit

　　钱培名出自钱圩钱氏一脉，曾在此校勘历代孤本名著，辑成《小万卷楼丛书》，清光绪年间辑成《钱氏汇刻书目》，收录金山钱氏历代所有汇刻书目。该宅现为朱鹏高艺术馆，展示朱鹏高多年的精品创作。

Qian Peiming, hailing from the Qianwei Village and the Qian family lineage, had conducted textual research on rare editions of famous works from various eras in this place. He compiled these works into the collection called "*The Collection of Xiao Wan Juan Lou*". During the Guangxu period of the Qing Dynasty, he further edited "*The Comprehensive Catalogue of the Qian Family's Republished Books*", which documented the collated books from the Qian family lineage throughout the generations. The residence has now been transformed into the Zhu Penggao Art Gallery, where it showcases the exquisite creations of Zhu Penggao over the years.

白蕉故居
Bai Jiao Former Residence

年代: 近现代
类别: 近现代重要史迹及代表性建筑
级别: 金山区文物保护点
利用情况: 居住场所
Era: Modern times
Category: Modern important historical sites and representative buildings
Conservation level: Jinshan district-level protected place
Utilization: Residence

　　白蕉（1907—1969），名馥，字远香，号旭如，笔名白蕉，张堰镇人，诗、书、画造诣较深，与徐悲鸿、邓散木并称"艺坛三杰"，1969年去世。张堰镇的白蕉故居有两处，分别位于新尚路及东大街。新尚路故居为传统的砖木结构、小青瓦平房，面宽5间，建筑面积147平方米，是白蕉出生及童年生活的地方。东大街故居由房屋4间及后花园组成，总占地面积185平方米，是白蕉青年时期居住的地方，白蕉成年后主要在此生活和创作书画。

　　Bai Jiao (1907–1969), originally named Fu, with the courtesy name Yuanxiang and the artistic names Xuru and Bai Jiao, was a prominent poet, calligrapher, and painter from Zhangyan Town. He was considered one of the "Three Masters of the Art World" along with Xu Beihong and Deng Sanmu. Bai Jiao passed away in 1969. There are two former residences of Bai Jiao in Zhangyan Town. One is located on Xinshang Road and the other on (East Street). The former residence on Xinshang Road is a traditional brick and wood structure with small gray tile roofs. It has a width of 5 rooms and a total area of 147 square meters. This was the place where Bai Jiao was born and spent his childhood. The former residence on Dongda Street consists of four rooms and a backyard, covering a total area of 185 square meters. This was the place where Bai Jiao lived during his youth. After reaching adulthood, he primarily resided here and created his literary and artistic works.

吴梁三命坊

Wu Liang San Ming Memorial Archway

年代：明代
类别：古建筑
级别：金山区文物保护点
利用情况：开放参观
Era: Ming Dynasty
Category: Ancient architecture
Conservation level: Jinshan district-level protected place
Utilization: Open for visit

    吴梁，生年不详，字百材，号贞石，张堰镇人。明嘉靖十六年（1537）举人，曾先后任福建郡武府推官、刑部侍郎、广西太平知府等职，先后发起筹建嘉庆桥、寿安桥、横桥、惠通桥、白带桥等。卒年87岁，卒后家乡民众为其建"三命坊"，以表彰其在为官经历中不畏强暴、为民请命的功德。据乾隆《金山县志》记载，清初清军攻金山卫城经张堰镇时坊被毁，现仅存南侧一根坊柱。

    Wu Liang, whose birth year is unknown, had the courtesy name Baicai and the artistic name Zhenshi. He was from Zhangyan Town. In the 16th year of the Jiajing reign of the Ming Dynasty (1537), he passed the imperial examination and held various positions such as the military judge of Fujian province, assistant minister of the Ministry of Justice, and the governor of Taiping in Guangxi province. He initiated the construction of bridges such as Jiaqing Bridge, Shou'an Bridge, Hengqiao Bridge, Huitong Bridge, and Baidai Bridge. He passed away at the age of 87. After his death, the local people in his hometown erected the San Ming Memorial Archway in his honor, to recognize his contributions in fearlessly standing up against tyranny and seeking justice for the people during his official career. According to the Qianlong edition of the "*Jinshan County Annals*", the monument was destroyed when the Qing army attacked the Jinshan Wei City through Zhangyan Town at the beginning of the Qing Dynasty. Only one pillar remains on the southern side now.

高桥镇

1 老宝山城遗址
高桥镇杨高北一路 285 号

2 高桥仰贤堂
高桥镇义王路 1 号

3 高桥黄氏民宅
高桥镇西街 139 号

4 钟氏民宅
高桥镇西街 160 号

5 凌氏民宅
高桥镇西街 167 号

6 蔡氏民宅
高桥镇季景北路 714 弄 11 号

1 *Old Baoshan City Ruins*
No.285, Yanggao North 1st Road, Gaoqiao Town

2 *Gaoqiao Yangxian Hall*
No.1, Yiwang Road, Gaoqiao Town

3 *Gaoqiao Huang Family Residence*
No.139, West Street, Gaoqiao Town

4 *Zhong Family Residence*
No. 160, West Street, Gaoqiao Town

5 *Ling Family Residence*
No. 167, West Street, Gaoqiao Town

6 *Cai Family Residence*
No.11, Lane 714, Jijing North Road, Gaoqiao Town

↑

① 老宝山城遗址
Old Baoshan City Ruins

② 高桥仰贤堂
Gaoqiao Yangxian Hall

⑥ 蔡氏民宅
Cai Family Residence

钟氏民宅
Zhong Family Residence

三峡石图艺术馆
Three Gorges Stone Picture Art Gallery

高桥黄氏民宅
Gaoqiao Huang Family Residence

⑤ 凌氏民宅
Ling Family Residence

钱慧安纪念馆
Qian Hui'an Memorial Hall

图例 LEGENDS

- **n** 全国重点文物保护单位
  National priority protected site
- **n** 上海市文物保护单位
  Shanghai city-level protected site
- **n** 区级文物保护单位
  District-level protected site
- **n** 区文物保护点
  District-level protected place
- ◎ 其他景点
  Other attractions
- ← 范围外景点
  Out-of-range attractions
- 历史风貌区范围
  Scope of historical district
- 游览路线
  Touring route

本图为位置示意，与实际尺寸不符
Illustration is not proportional to the actual scale

高桥镇
Gaoqiao Town

交通指南：
地铁 6 号线至外高桥保税区站，步行 15 分钟可达；
地铁 10 号线至高桥站，换乘公交 508 路至市七医
院站，步行 5 分钟可达。

Transportation Guide:
Metro Line 6 to Waigaoqiao Free Trade Zone
Station, 15 minutes' walk;
Metro Line 10 to Gaogaoqiao Station, transfer
to Bus No. 508 to Shanghai Seventh People's
Hospital Station, 5 minutes' walk.

海防重镇，营造之乡

高桥，因桥而得名，史称清溪，位于浦东新区北隅，长江、黄浦江和东海的"三水"交汇处，有着"万里长江口，千年高桥镇"的美誉。

高桥成陆于唐，南宋建炎三年（1129）始建临江乡，为高桥建置之始。宋元时期，高桥盐业兴盛。先民的聚居地为清浦里，也叫清浦场，诞生了高桥东部、长江口以南的第一个盐场。元代，高桥航海业开始兴盛。元初高桥人张瑄为元帝开通大规模海上漕运，营造大批平底海船即"沙船"。当地不少百姓参与沙船运输业，开拓近海贸易。明清时期的高桥为海防重地，洪武年间设立清浦旱寨，至明代成化、弘治年间已经是江东重镇，寺庙鼎立，市镇繁荣。彼时，高桥镇纺织业兴盛，出现了品牌化和规模化。据《宝山县续志》载，明、清两代由沙船运往北方牛庄等地的纱布皆出自高桥。

上海开埠以后，高桥得西方风气之先，建筑营造技艺得到极大发展，"一把泥刀"红遍上海，走向全国。营造爱俪园以及和平饭店的王松云，被誉为"上海石库门文化奠基人"的钟惠山，修建俄罗斯联邦驻上海总领事馆的周瑞庭，以及承包麦边洋行、怡和洋行、天祥洋行三大工程的谢秉衡等人皆出自高桥。他们不仅在全国营造了许多近代著名的建筑，还在自己的家乡建造了结构精良、中西合璧的私人宅第。此外，高桥乡民也多从事餐饮、刺绣等行业，高桥也被誉为浦东"三刀一针"（泥刀、菜刀、剪刀以及绣花针）的发祥地。

高桥镇历史格局较为完整，镇内保存有"丁"字形两条干河和东、西、北三条傍河而建的长街，三条老街又构成"丁"字形骨架，体现三水交汇区域河网密布的自然城镇特色。历史建筑主要集中在"丁"字形街道周边，类型丰富，传统建筑与中西合璧的海派民居融为一体，与各种水埠、码头和桥梁共同构成了高桥"河街共生、水上人家"的江南水乡风光。除了物质遗存，高桥还保留了数量众多的非物质文化遗产，如国家级非遗"上海绒绣"、市级非遗"高桥松饼制作技艺""本帮菜传统烹饪技艺"等，大多依托高桥"三刀一针"的传统技艺而来。

从渔盐之乡、沙船之乡、纺织之乡到营造之乡，高桥镇在千百年的历史发展中孕育了丰富的乡土文化，凝结成了为数众多的文化遗产。其"因桥而名"的命名特色、中西合璧的营造特色和海防海运历史文化特色充分展现了浦东地区江南文化和海派文化的交融与传承，是上海古今变迁的见证。

不可移动文物资源

2010 年 7 月，高桥镇被公布为第五批中国历史文化名镇，其中历史文化风貌保护区面积 56.3 公顷，核心保护区面积 9.4 公顷。镇域内现有各级不可移动文物 75 处，包括上海市文物保护单位 3 处，区级文物保护单位 7 处，区文物保护点 65 处。类型以民居为主，海派风格显著。

高桥镇参观指南 Gaoqiao Town Visiting Guide

游览路线：
高桥仰贤堂（高桥历史文化陈列馆）→三峡石图艺术馆→高桥黄氏民宅（高桥绒绣馆）→凌氏民宅（高桥人家陈列馆）→钱慧安纪念馆

Tourist Route:
Gaoqiao Yangxian Hall (Gaoqiao History and Culture Exhibition Hall) → Three Gorges Stone Picture Art Gallery → Gaoqiao Huang Family Residence (Gaoqiao Woolen Embroidery Museum) → Ling Family Residence (Gaoqiao Family Exhibition Hall) → Qian Hui'an Memorial Hall

古镇美食特产：
高桥松饼

Local Specialties:
Gaoqiao Muffin

A town of coastal defense, known for its architecture

Gaoqiao, named after the bridge, is also known as Qingxi. It is located in the northern corner of Pudong New Area, at the confluence of the Yangtze River, Huangpu River, and East China Sea. It is renowned as the "The mouth of the Yangtze River for ten thousands li, the town of Gaoqiao for one thousand of year".

Gaoqiao became a land in the Tang Dynasty, and in the third year of the Jianyan period in the Southern Song Dynasty (1129), the village of Linjiang was established, marking the beginning of Gaoqiao's establishment. During the Song and Yuan Dynasties, salt production prospered in Gaoqiao. The original settlement of the ancestors was called Qingpu Village, which gave birth to the first salt field east of Gaoqiao and south of the Yangtze River mouth. During the Yuan Dynasty, Gaoqiao's maritime industry began to flourish. Zhang Xuan, a native of Gaoqiao, opened up large-scale maritime transportation for the Yuan Emperor, creating a large number of flat-bottomed sea ships called "sand boat". Many locals participated in the sand boat transportation business and developed coastal trade. In the Ming and Qing Dynasties, Gaoqiao was an important coastal defense area. During the Hongwu period, the Qingpu Dry Barracks were established, and by the Chenghua and Hongzhi periods in the Ming Dynasty, it had become a major town in Jiangdong with numerous temples and a prosperous market. During this period, the textile industry in Gaoqiao flourished, with the emergence of branding and scaling. According to the "Continuation of the Annals of Baoshan County", the muslin cloth transported by sand boats to places like Niu Zhuang in the north during the Ming and Qing Dynasties all came from Gaoqiao.

After the opening of Shanghai, Gaoqiao was influenced by Western culture, and its construction and craftsmanship developed rapidly. The "One Mud Knife" technique became well-known in Shanghai and spread throughout the country. Wang Songyun, who built the Aaron & Liza Garden and the Peace Hotel, Zhong Huishan, known as the "founder of Shanghai's Shikumen culture", Zhou Ruiting, who built the Consulate General of the Russian Federation in Shanghai, and Xie Bingheng, who contracted the McBain & Co., Jardine Matheson & Co., and Adamson & Co., all came from Gaoqiao. They not only built many famous modern buildings nationwide but also built well-structured, fusion of Chinese and Western private residences in their hometown. In addition, many Gaoqiao villagers engage in the catering and embroidery industries, earning Gaoqiao the reputation as the birthplace of the "Three Knives and One Needle" (mud knife, kitchen knife, scissors, and embroidery needle) in Pudong.

Gaoqiao Town has a relatively complete historical layout. It has two main canals in the shape of the Chinese character " 丁 " (ding), with three long streets built along the east, west, and north sides of the canals. These three old streets form the framework of the " 丁 " shape, showcasing the natural urban characteristics of the region where three waterways converge and create a densely woven river network. The historical buildings in Gaoqiao Town are mainly concentrated around the streets in the shape of the Chinese character " 丁 " . These buildings are diverse in their architectural styles, blending traditional Chinese architecture with Western influences, creating a unique blend that is characteristic of the Shanghai style. These historical buildings, combined with various waterfronts, docks, and bridges, create the picturesque scenery of Gaoqiao as a "river and street coexistence, waterborne households" in the Jiangnan water town style. In addition to the physical heritage, Gaoqiao also preserves a large number of intangible cultural heritage. For example, "Shanghai Woolen Embroidery" is a national-level intangible cultural heritage, while "Gaoqiao Muffin Making Technique" and "Shanghai Local Cuisine Traditional Cooking Techniques" are both city-level intangible cultural heritages. These cultural heritages are mostly based on the traditional craftsmanship of "three knives and one needle" in Gaoqiao.

From being known as the hometown of fishing and salt production, to being recognized as a center for shipbuilding and textile production, and now being recognized as a town known for its construction industry, Gaoqiao Town has nurtured a rich local culture over thousands of years of historical development. This has resulted in a plethora of cultural heritage. Gaoqiao's distinctive naming tradition of being named after bridges, its blend of Chinese and Western architectural styles, and its historical significance in maritime defense and transportation all showcase the integration and inheritance of Jiangnan and Shanghai-style culture in the Pudong area. Gaoqiao is a witness to the historical changes and development of Shanghai throughout the ages.

Immovable Cultural Relics Resources

In July 2010, Gaoqiao Town was announced as the fifth batch of national famous historical and cultural towns. The historical and cultural preservation area covers an area of 56.3 hectares, with a core protection area of 9.4 hectares. There are currently 75 immovable cultural relics in the town, including 3 Shanghai city-level protected sites, 7 district-level protected sites and 65 district-level protected places. Most of these cultural relics are residential buildings with a significant Shanghai-style architectural influence.

老宝山城遗址
Old Baoshan City Ruins

年代： 清代
类别： 古遗址
保护级别： 上海市文物保护单位
利用情况： 其他
Era: Qing Dynasty
Category: Ancient ruins
Conservation level: Shanghai city-level protected site
Utilization: Other uses

　　老宝山城筑于清康熙三十三年（1694），占地4万余平方米，方形，设4门，纵横十字街，现存城门洞及城墙基础一段，是高桥历史上最大的军事堡垒，也是吴淞江入海口的屏障。清雍正元年（1723），分嘉定县东北四乡设宝山县，县治设于吴淞所，即今宝山区政府所在地，称宝山县城，位于高桥的宝山城遂称老宝山城。老宝山城是目前上海少有的用于屯兵的古城遗址之一。

　　The Old Baoshan City ruins were built in the 33rd year of the Kangxi reign of the Qing Dynasty (1694). It covered an area of over 40 000 square meters and had a square layout with four gates and a crisscrossing network of streets. The remains of the city gate and the foundation of the city wall are still preserved. It was the largest military fortress in the history of Gaoqiao and served as a barrier at the mouth of the Wusong River into the sea. In the 1st year of the Yongzheng reign of the Qing Dynasty (1723), Baoshan County was established, covering four towns northeast of Jiading County. The county seat was located at Wusong, which is now the location of the Baoshan District government. The city in Baoshan was hence named the Old Baoshan City. The Old Baoshan City is one of the few ancient city ruins in Shanghai that were used for garrison purposes.

高桥仰贤堂
Gaoqiao Yangxian Hall

年代: 近现代
类别: 近现代重要史迹及代表性建筑
保护级别: 上海市文物保护单位
利用情况: 开放参观
Era: Modern times
Category: Modern important historical sites and representative buildings
Conservation level: Shanghai city-level protected site
Utilization: Open for visit

原为沈晋福旧居,建于 20 世纪 30 年代,1933 年竣工。仰贤堂坐北朝南,建筑为中西合璧的二层砖木结构,正面看是中式宅院,从背面隔河相望,又像是一座建在水上的西式别墅。平顶为晒台,楼下有地下室,是高桥镇上第一家建有地下室的住宅。四周有 5 米高的封火墙,沿河建有坝岸和两座河埠,楼上沿河设阳台。内部装修富丽堂皇,不仅镶嵌着精致的木雕和字画,也安装着西式的壁炉吊灯,做工用料精益求精,沿河驳岸工程和地下室防水层至今牢固如初。现为"高桥历史文化陈列馆",展现高桥的历史文化、生产生活以及乡风民俗等。

Gaoqiao Yangxian Hall, originally known as the former residence of Shen Jinfu, was built in the 1930s and completed in 1933. Facing south, the Yangxian Hall is a two-story brick and wood structure that combines Chinese and Western architectural styles. From the front, it appears as a traditional Chinese courtyard, while from the back, it resembles a Western villa seemingly built on the water. The flat roof serves as a sun deck, and there is a basement underneath, making it the first residential building in Gaoqiao Town to have an underground space. Surrounding the building is a 5-meter-high firewall, and along the river, there are embankments and two river ports. There are balconies on the upper floor facing the river. The interior is elaborately decorated, featuring exquisite wood carvings, calligraphy, and paintings, as well as Western-style fireplace and chandeliers. The construction and materials used are of high quality, and the embankment and waterproof layer of the basement remain strong and intact. It is currently functioning as the "Gaoqiao Historical and Cultural Exhibition Hall", showcasing various aspects of Gaoqiao's history, culture, production, lifestyle, and local customs.

高桥黄氏民宅

Gaoqiao Huang Family Residence

年代: 近现代
类别: 近现代重要史迹及代表性建筑
保护级别: 浦东新区文物保护单位
利用情况: 开放参观
Era: Modern times
Category: Modern important historical sites and representative buildings
Conservation level: Pudong New Area district-level protected site
Utilization: Open for visit

　　始建于 20 世纪初，建筑面积约 1000 平方米。坐北朝南，前有砖雕门楼。三进，砖木结构，一、二进为二层楼房，底层船篷轩回廊，二层走马楼，堂挂"润德堂"匾额，梁枋、窗棂、门扇多有精致木雕。第三进有 5 间平房。宅主黄文钦，生于 1858 年，早年在上海苏州河畔的乌镇路、栗阳路等处开设数家竹行，俗称为"黄家竹行"。该宅有大小房间共 25 间，至 1921 年才全部完工。现开辟为"高桥绒绣馆"。

Gaoqiao Huang Family Residence, originally built in the early 20th century, has a total floor area of approximately 1000 square meters. It faces south, with a brick-carved gatehouse at the front. The residence consists of three sections and is built with a combination of brick and wood structure. The first and second sections are two-story buildings, with a lower-level veranda and an upper-level Zou Ma Lou style corridor. The hall is adorned with the plaque "Runde Hall", and the beams, window frames, and doors feature exquisite wood carvings. The third section comprises five single-story houses. The owner of the residence, Huang Wenqin, was born in 1858. In his early years, he operated several bamboo businesses in areas such as Wuzhen Road and Liyang Road on the Suzhou River in Shanghai, known as the "Huang Family Bamboo Business". The residence consists of a total of 25 rooms and was completed in 1921. It has now been converted into the "Gaoqiao Woolen Embroidery Museum".

钟氏民宅
Zhong Family Residence

年代: 近现代
类别: 近现代重要史迹及代表性建筑
保护级别: 浦东新区文物保护单位
利用情况: 商业用途
Era: Modern times
Category: Modern important historical sites and representative buildings
Conservation level: Pudong New Area district-level protected site
Utilization: Commercial use

钟氏民宅为中西合璧的砖木结构二层楼。砖墙立柱，抬梁式混合结构。水泥粉刷外墙，红瓦坡顶。前为歇山顶门楼，门额砖雕刻"竹苞松茂"四字。在南北向主轴线上建正厅、正宅、内室，旁有附房。20世纪20年代末，由当时的"高桥首富"、被誉为"石库门大王"的著名建筑营造商钟惠山投资20万银元建造。现开辟为集咖啡、美食、文创等于一体的一尺花园。

The Zhong Family Residence is a two-story brick and wood structure that combines Chinese and Western architectural elements. It has brick walls, columns, and a column and strut-beam construction. The exterior walls are coated with cement, and it features a sloping roof with red tiles. At the front, there is a gatehouse with a gable and hip roof, and the horizontal inscribed brick over it carves the phrase "bamboo buds and flourishing pines". The main axis of the residence runs from north to south and includes the main hall, main house, inner rooms, and adjacent rooms. In the late 1920s, it was built with an investment of 200 000 silver dollars by Zhong Huishan, a famous builder who was known as the "King of Shikumen" and hailed as the "wealthiest person in Gaoqiao" at the time. It has now been transformed into a one-foot garden that includes a cafe, gourmet food, and cultural and creative products.

凌氏民宅
Ling Family Residence

年代: 近现代
类别: 近现代重要史迹及代表性建筑
保护级别: 浦东新区文物保护点
利用情况: 开放参观
Era: Modern times
Category: Modern important historical sites and representative buildings
Conservation level: Pudong New Area district-level protected place
Utilization: Open for visit

　　凌氏民宅又名三德堂，建于 1918 年，主人为凌祥春，曾为高桥富户。该宅二层砖木结构，三进院落，二层建回廊，前后各有庭院。主入口朝北，从西街进入，南侧临河有后院，院墙上设有院门两座。正厅廊步皆有轩，隔扇门窗镶嵌玻璃。在门楼的装饰上采用了当时新式的水泥堆塑，题材为花草。该宅规模较大，雕饰精美，是当地民居的典型代表，现为"高桥人家陈列馆"，展示民国时期高桥地区四世同堂大家庭的生活场景。

　　The Ling Family Residence, also known as the Sande Hall, was built in 1918 and owned by Ling Xiangchun, a wealthy resident of Gaoqiao. The residence is a two-story brick and wood structure with three courtyards. It has a second-floor corridor and front and back gardens. The main entrance faces north and is accessed from the western street. There is a backyard on the south side, adjacent to the river, with two gates on the courtyard walls. The main hall and corridors are all adorned with verandas, and the doors and windows are fitted with glass insets. The decoration of the gatehouse utilizes a new form of cement sculpture at the time, depicting flowers and plants. The residence is of a large scale and features exquisite carvings, making it a typical representative of local residences. It is now the "Gaoqiao Family Exhibition Hall", showcasing the living scenes of large extended families in Gaoqiao during the Republican era.

蔡氏民宅
Cai Family Residence

年代: 清代
类别: 近现代重要史迹及代表性建筑
保护级别: 浦东新区文物保护点
利用情况: 居住场所
Era: Qing Dynasty
Category: Modern important historical sites and representative buildings
Conservation level: Pudong New Area district-level protected place
Utilization: Residence

　　蔡氏民宅建成于清光绪三十四年（1908），原宅主蔡啸松，因经营米业成为高桥巨富。该宅坐西朝东，三开间、三进院落，二层砖木结构，青砖青瓦白灰墙，墙门檐装饰精美砖雕。客堂正上方有"庆誉堂"三字匾额。其用材和做工上都属上乘，耗银4000余两，历时达13年之久才落成。其规模宏大、雕饰精美，是当地民居的杰出代表。

The Cai Family Residence, built in the 34th year of the Guangxu reign of the Qing Dynasty (1908), was originally owned by Cai Xiaosong, who became a wealthy figure in Gaoqiao through his rice business. The residence faces east, with three bays and three courtyards. It is a two-story brick and wood structure, with green bricks, green tiles, and white plaster walls. The walls and door eaves are exquisitely decorated with brick carvings. Above the main hall, there is a plaque with the three characters "Qingyu Hall". The materials and craftsmanship of the residence are of the highest quality, costing more than 4000 taels of silver and taking 13 years to complete. The residence is of grand scale and exquisite carvings, making it an outstanding representative of local residences.

金泽镇	Jinze Town
1 普济桥 金泽镇南段	**1** *Puji Bridge* South section of Jinze Town
2 迎祥桥 金泽镇南段	**2** *Yingxiang Bridge* South section of Jinze Town
3 万安桥 金泽镇北段	**3** *Wan'an Bridge* North section of Jinze Town
4 林老桥 金泽镇北段	**4** *Linlao Bridge* North section of Jinze Town
5 如意桥 金泽镇南段	**5** *Ruyi Bridge* South section of Jinze Town
6 天皇阁桥 金泽镇北段	**6** *Tianhuangge Bridge* North section of Jinze Town
7 金泽放生桥 金泽镇南段	**7** *Jinze Fangsheng Bridge* South section of Jinze Town
8 颐浩寺遗址 金泽镇迎祥街 12 号	**8** *Yihao Temple Ruins* No.12, Yingxiang Street, Jinze Town

图例 LEGENDS

全国重点文物保护单位
National priority protected site

上海市文物保护单位
Shanghai city-level protected site

区级文物保护单位
District-level protected site

区文物保护点
District-level protected place

其他景点
Other attractions

历史风貌区范围
Scope of historical disctrict

游览路线
Touring route

林老桥
Linlao Bridge

万安桥
Wan'an Bridge

天皇阁桥
Tianhuangge Bridge

北胜浜街 Beishengbang Street

北胜浜 Beishengbang River

金溪路 Jinxi Road

颐浩寺遗址
Yihao Temple Ruins

迎祥街 Yingxiang Street

陈道浜 Chendaobang River

上塘街 Shangtang Street

金泽塘 Jinze Pond

下塘街 Xiatang Street

普济桥
Puji Bridge

金泽放生桥
Jinze Fangsheng Bridge

如意桥
Ruyi Bridge

东库港 Dongku Port

迎祥桥
Yingxiang Bridge

金溪路 Jinxi Road

本图为位置示意，与实际尺寸不符
Illustration is not proportional to the actual scale

金泽镇

Jinze Town

交通指南:
沪商高速专线至金泽汽车站，步行 500 米可达。

Transportation Guide:
Bus Hushang High-speed Dedicated Line to Jinze
Bus Station, within 500 meters walking distance.

原生活态，江南桥乡

凡到金泽，首先体会的是江南水乡的宁静恬和，清澄明澈的市河水，洁净如洗的石板路，古老质朴的石拱桥，临河两岸粉墙黛瓦的院落连绵错落，犹如一组组水墨山水。河岸边原住民轻摇蒲扇，巷弄里饭菜飘香，这里原真地承载了活态的水乡风貌，令人流连忘返。

金泽，位于今青浦区西南，因在白米港畔，盛产大米、棉花、苎麻，又是运米的聚集之地，古称"白苎里"。后又名"金溪"，诗云"金溪面上水滢洄"，就是形容此地水多。也因多湖荡水泽，方便对境内田块灌溉，人们引"穑人获泽如金"之句，定镇名为"金泽"。金泽也确实以多水著称，北枕淀山湖，南依太浦河，处"九峰三泖"之地，江河湖港交织，水网密布，水域面积占全镇总面积的三分之一以上，是闻名江南的水乡泽国。其历史悠久，4000多年前的良渚时期，就有先民在此活动。唐末至五代时，北方战乱频频，江南一带较为安定，战乱中的人们南迁至此，发现白苎里土地肥沃、气候温和，遂安家垦殖，由此人丁日众。南宋初年宰相吕颐浩在金泽建造府第，由此也奠定了古镇规模。后吕颐浩舍宅为寺，成就了恢宏大气的江南名刹颐浩寺。至元代愈加繁盛，明代棉纺织业兴盛，商业发达，清代已发展成"居民数千家"的大镇。清《金泽小志》记载有"六观、一塔、十三坊、四十二虹桥，桥各有庙"，市镇之繁华，古桥数量之多，寺庙之多，冠绝江南。

金泽镇区中心地势高隆，镇域四周分布着大小圩头，地形似龟，灵秀天成，民间盛传"地如金龟、风水福泽"，因此有"金龟福地、鱼米之乡"之誉。因势制宜，古镇以金泽塘及北胜浜为水路主干，其余为支流，构成"一纵一横多分支"的水系脉络。如今仍较为完整地保存了旧时格局和风貌，河道间分布着众多古桥驳岸，联系贯通起街巷与宅院，条石铺就的上、下塘街，保持着明清风格，街道巷弄古意悠然，临河民居鳞次栉比，宅院屋舍内别有洞天，桥头古刹畔香火缭绕，这里完整而真实地保存了水乡古镇的形态，堪称江南古镇的活化石。

金龟福地的传说和星罗棋布的古桥造就了金泽"桥桥有庙、庙庙有桥"的桥庙文化，是金泽独特的人文景观。虽历经时代变革，庙宇多数不存，但镇上仍随处可寻共存的桥庙文化空间。放生桥头的总管庙、林老桥畔的关帝庙、桥庙相依，天皇阁桥旁的天王庙、如意桥旁的祖师庙、迎祥桥畔的万寿庵，史志中仍有迹可循。

桥庙文化孕育了金泽独具特色的民间信仰与文化遗存，衍生出民俗、饮食、演艺等丰富多彩的地方文化。金泽拥有丰富的非物质文化遗产，有国家级非物质文化遗产1项，即吴歌（青浦田山歌），上海市非物质文化遗产5项，分别为阿婆茶、宣卷、烙画、箍具制作技艺、金泽庙会，区级非物质文化遗产2项，分别为赵家豆腐干制作技艺、金泽状元糕制作技艺。各类非物质文化遗产各具特色，田山歌是一种口头传唱形式的原始演唱艺术，反映了田间劳作民众的现实生活，在金泽四乡已有数百年历史。上海市非遗阿婆茶作为"以茶为礼、以茶待客"的风俗礼仪，在金泽地区流传甚广。上海市非遗宣卷，传承自唐代"俗讲"和宋代"谈经"的说唱艺术，颇具江南水乡韵味。上海市级非遗烙画把绘画艺术的各

金泽镇参观指南 Jinze Town Visiting Guide

游览路线：
林老桥→万安桥→天皇阁桥→普济桥→颐浩禅寺→放生桥→如意桥→迎祥桥

古镇美食特产：
金泽状元糕、赵家豆腐干、梅花糖豆

Tourist Route:
Linlao Bridge → Wan'an Bridge → Tianhuangge Bridge → Puji Bridge → Yihao Zen Temple → Fangsheng Bridge → Ruyi Bridge → Yingxiang Bridge

Local Specialties:
Jinze Zhuangyuan Cake, Zhao Family Dried Bean Curd, Plum Blossom Sugar Beans

种表现技法与烙画艺术融为一体，作品古朴典雅，精美绝伦。

不可移动文物资源

2014 年 2 月，金泽被公布为第六批中国历史文化名镇，其中历史文化风貌保护区面积 51.84 公顷，核心保护区面积 17.94 公顷。金泽文物众多，镇域内现有各级不可移动文物 61 处，包括上海市文物保护单位 2 处，区级文物保护单位 6 处，区文物保护点 53 处。在金泽塘上不足 400 米的空间里就集聚了普济桥、迎祥桥、放生桥和如意桥等跨越宋、元、明、清四个朝代的古桥梁，分布之密集、形态之丰富、时代跨度之久远，为其他古镇所无法比拟，故被誉为"古桥梁博物馆"，是名副其实的江南桥乡。

A town of bridges, with original Jiangnan lifestyle

Upon arriving in Jinze, the first thing you will experience is the tranquility and serenity of a water town in Jiangnan. The clear and pristine waters of the city's river, the immaculate stone-paved roads, the ancient and rustic stone arch bridges, and the rows of white-walled, black-tiled courtyard houses along the riverbanks, all resemble ink-washed landscapes. Native residents gently fan themselves on the riverbank, while delicious aromas of food waft through the narrow alleyways. This place truly embodies the vibrant charm of a water town, leaving visitors lingering and reluctant to leave.

Jinze is located in the southwest of Qingpu district, seated banks of the Baimi Port, is known for its abundant production of rice, cotton, and ramie, as well as its role as a gathering place for rice transportation. It was originally called "Bai Zhu Li" and later renamed "Jinxi" due to its abundance of water. It is described in a poem as "the shimmering water on the surface of Jinxi". The town is also known for its numerous lakes, which make irrigation of the farmland convenient. The phrase "harvesting water is as precious as gold" inspired the name "Jinze". Jinze is indeed famous for its abundance of water. It is bordered to the north by Dianshan Lake and to the south by the Taipu River, situated in a region surrounded by nine peaks and three ponds, with rivers, lakes, and ports interlacing with each other. The water area accounts for more than one-third of the total area of the town, making it a well-known water town in the Jiangnan region. The town has a long history. More than 4000 years ago during the Liangzhu period, ancient people were already active in this area. During the late Tang Dynasty and the Five Dynasties period, frequent wars occurred in the northern regions while the Jiangnan area remained relatively stable. People fleeing the wars migrated to this area and found the fertile land and mild climate of Bai Zhu Li exceptional for settlement and cultivation. This led to a growth in population. In the early years of the Southern Song Dynasty, Prime Minister Lü Yihao built his residence in Jinze, which laid the foundation for the development of the ancient town. Later, Lü Yihao devoted his residence to be converted into a temple, giving rise to the magnificent Yihao Zen Temple, a famous Buddhist temple in Jiangnan. During the Yuan Dynasty, the town further prospered, and in the Ming Dynasty, the cotton textile industry flourished, leading to a developed commercial sector. By the Qing Dynasty, Jinze had already become a prosperous town with thousands of residents. The Qing Dynasty's "Brief History of Jinze" recorded "six temples, one pagoda, thirteen neighborhoods, and forty-two rainbow bridges, each bridge with its own temple". The town's prosperity, the number of ancient bridges, and the abundance of temples are unparalleled in the Jiangnan region.

Jinze Town is located in a high and prominent area, surrounded by numerous small and large weirs (traditional agricultural water facilities). The terrain resembles a turtle, with its natural beauty and charm often described by locals as a "fortunate land with the blessings of Fengshui". It is known as the "Golden turtle paradise and the land of fish and rice". Adapting to the natural conditions, the ancient town of Jinze has positioned Jinze Pond and Beishengbang River as the main waterways, while the rest serve

as tributaries, forming a water system with "one vertical, one horizontal, and multiple branches". Today, the old layout and style have been well-preserved. There are numerous ancient bridges and embankments distributed between the river channels, connecting streets, alleys, and residential courtyards. Shangtang and Xiatang streets, paved with flagstones, maintain a style reminiscent of the Ming and Qing dynasties. The streets and lanes exude a sense of tranquility and ancient charm, while the homes along the river are densely packed, each with unique characteristics. Alongside the ancient temples near the bridges, incense continuously wafts through the air. This place has successfully preserved the form of a typical water town, making it a living fossil of ancient towns in the southern region.

The legend of "Golden Turtle Paradise"

and the numerous ancient bridges scattered throughout Jinze have created the unique bridge and temple culture of Jinze, known as "bridges beside temples, temples beside bridges". Despite the changes of time, most of the temples no longer exist, but the town still retains the cultural spaces where bridges and temples coexist. The Zongguan Temple at Fangsheng Bridge, the Guandi Temple by Linlao Bridge, the Bridge Temples in close proximity, the Tianwang Temple next to Tianhuangge Bridge, the Zushi Temple beside Ruyi Bridge, the Wanshou Temple by Yingxiang Bridge can still be found in historical records.

The cultural heritage of the Bridge Temples has nurtured distinctive folk beliefs and cultural relics in Jinze, giving rise to a rich and diverse local culture, including customs, cuisine, and performing arts. Jinze boasts a rich intangible cultural heritage, including one national-level intangible cultural heritage item, the Wu Ge (Farm Song of Qingpu), five Shanghai municipal-level intangible cultural heritage items: Apo Tea (a traditional tea drinking customs), Xuanjuan (a traditional form of Chinese painting), Laohua (Pyrography), Duanju Furniture Making Techniques, and Jinze Temple Fair, as well as two district-level intangible cultural heritage items: Zhao Family Dried Bean Curd Production Techniques and Jinze Zhuangyuan Cake Making Techniques. Various intangible cultural heritages have their own characteristics. Farm Song, is a primitive vocal art that has been passed down orally and reflects the real life of field laborers. It has a history of several hundred years in the four townships of Jinze area. The custom and etiquette of "using tea as a gift and serving guests with tea" of the Shanghai municipal-level intangible cultural heritage, Apo Tea, is a folk custom spreads in the Jinze area. Xuanjuan, a Shanghai municipal intangible cultural heritage, is a storytelling and singing art that originated from the Tang Dynasty's "Sujiang" (the chanting and storytelling art form depicting the lives of common people, which was typically performed and spread by folk artists through oral tradition) and the Song Dynasty's "Tanjing" (This form of chanting and storytelling was based on Confucian classics and aimed to educate and guide the audience through discussing the content of these classics, as well as exploring moral and ethical aspects of life). It exudes the charm of the water towns in Jiangnan region. Shanghai municipal intangible cultural heritage Lao Hua integrates various painting techniques with pyrography art, creating works that are elegant and exquisite.

Immovable cultural relics resources

In February 2014, Jinze was announced as the sixth batch of national famous historical and cultural towns. The historical and cultural preservation area covers an area of 51.84 hectares, with a core protected area of 17.94 hectares. Jinze is rich in cultural heritage, and there are currently 61 immovable cultural relics within the town, including 3 Shanghai city-level protected sites, 6 district-level protected sites and 53 district-level protected places. Within a space of less than 400 meters along Jinze Pond, there are ancient bridges spanning the Song, Yuan, Ming, and Qing dynasties, such as Puji Bridge, Yingxiang Bridge, Fangsheng Bridge, and Ruyi Bridge. The density of their distribution, the richness of their forms, and their long historical span are unparalleled by other ancient towns, earning the reputation of being a "museum of ancient bridges" and a true bridge village in the Jiangnan region.

普济桥
Puji Bridge

年代: 宋代
类别: 古建筑
保护级别: 上海市文物保护单位
利用情况: 开放参观
Era: Song Dynasty
Category: Ancient architecture
Conservation level: Shanghai city-level protected site
Utilization: Open for visit

始建于宋咸淳三年（1267），清雍正初重修，又称"圣堂桥"，因桥身材质为武康石，呈紫色，俗称"紫石桥"。系单孔石拱桥，东西向，跨金泽塘，桥长 25.5 米，宽 2.7 米，跨径 10.5 米，拱券上镌有"咸淳三年建"题款，桥面两侧有低矮的石护栏，东堍为 T 字形。普济桥弧形单孔，桥身较窄，坡度平缓，具有宋代桥梁特点，是上海地区最古老、保存最完好的古桥之一，有"上海第一古桥"的美誉。

Puji Bridge, also known as Shengtang Bridge, was first built in the third year of the Xianchun reign of the Song Dynasty (1267) and was later repaired during the early years of the Qing Dynasty's Yongzheng reign. It is also known as the "Purple Stone Bridge" due to its purple-colored Wukang stone material. Puji Bridge is a single-arch stone bridge, running in an east-west direction and crossing the Jinze Pond. It has a length of 25.5 meters, a width of 2.7 meters, and a span of 10.5 meters. The bridge arch bears the inscription "Built in the 3rd year of the Xianchun reign". There are low stone railings on both sides of the bridge, and the eastern end of the bridge is in the shape of a "T". Puji Bridge is a single-arched bridge with a narrow bridge deck and a gentle slope. It exhibits typical characteristics of bridges from the Song Dynasty and is one of the oldest and best-preserved ancient bridges in the Shanghai area, earning the reputation of being the "first ancient bridge of Shanghai".

迎祥桥
Yingxiang Bridge

年代：明代
类别：古建筑
保护级别：上海市文物保护单位
利用情况：开放参观
Era: Ming Dynasty
Category: Ancient architecture
Conservation level: Shanghai city-level protected site
Utilization: Open for visit

始建于元惠宗至元年间（1335—1340），明天顺六年（1462）重建，清乾隆三十三年（1768）重修，桥西原为元代有江南"小天竺"之称的万寿庵。迎祥桥为五跨简支梁石墩式桥，东西向，跨金泽塘，桥长 33 米，宽 2.45 米，桥身为砖、木、石混合结构，桥面呈弧形，由青砖铺砌，桥柱用青石拼成，石柱架条石作为横梁，横梁上密排楠木，选材独特，结构精巧，被誉为"连续简支"梁桥的鼻祖。"迎祥夜月"为金泽古八景之一，可见"长虹横卧、月印川流、水天一色"，景色宜人。

The Yingxiang Bridge, originally constructed during the Yuan Dynasty's Zhiyuan era (Huizong's reign, 1335–1340), was rebuilt in the 6th year of the Ming Dynasty's Tianshun reign (1462) and later underwent renovations in the 33rd year of the Qing Dynasty's Qianlong reign (1768). To the west of the bridge, there used to be a temple called Wanshou An, also known as the "Little Tianzhu" of Jiangnan during the Yuan Dynasty. The Yingxiang Bridge is a five-span simply supported stone pier bridge that runs east to west. It spans over the Jinze Pond and measures 33 meters in length and 2.45 meters in width. The bridge is a mixture of brick, wood, and stone structure, with an arched bridge deck paved with gray bricks. The bridge columns are made of blue stones, and the stone columns support the crossbeams, which are tightly spaced with nanmu wood. The selection of materials and the delicate structure make it the precursor of the "continuous simply supported beam" bridge. The "Yingxiang Night Moon" is one of the ancient Eight Scenic Views of Jinze, presenting a picturesque landscape of a rainbow lying across, the moon reflecting on the flowing river, and the harmony of water and sky.

万安桥
Wan'an Bridge

年代: 宋代
类别: 古建筑
保护级别: 青浦区文物保护单位
利用情况: 开放参观
Era: Song Dynasty
Category: Ancient architecture
Conservation level: Qingpu district-level protected site
Utilization: Open for visit

　　始建于宋景定年间（1260—1264），元至正二年（1342）桥上加建廊亭，又称"万安亭桥"。明嘉靖四十年（1561）、万历四十八年（1620）、清乾隆五十六年（1791）重修。系单孔石拱桥，东西向，跨金泽塘，桥长32.4米，宽3.1米，跨径10.2米，紫色武康石材质。桥面两侧有低矮的石护栏，雕水云纹，平面略呈弧形，坡度平缓。万安桥的结构、造型、用材与普济桥相似，两桥同跨一河，南北相峙，被称为"姊妹桥"。旧有"金泽四十二虹桥，万安为首"之说。

　　Wan'an Bridge was originally built during the Jingding period of the Song Dynasty (1260–1264). In the 2nd year of the Zhizheng reign of the Yuan Dynasty (1342), a pavilion was added to the bridge, hence it is also known as "Wan'an Ting Bridge". It underwent renovations in the 40th year of the Jiajing reign (1561), the 48th year of the Wanli reign (1620), and the 56th year of the Qianlong reign (1791). The bridge is a single-arch stone bridge, running from east to west, spanning the Jinze Pond. Made of purple Wukang stone, it is 32.4 meters long, 3.1 meters wide, with a span of 10.2 meters. There are low stone guardrails with water and cloud patterns on both sides of the bridge, with a slightly curved plane and a gentle slope. The structure, design, and materials of Wan'an Bridge are similar to those of Puji Bridge. They both span the same river, facing each other from the north and south, and are called "sister bridges". It is said that there were once "42 rainbow bridges in Jinze, with Wan'an as the first".

林老桥
Linlao Bridge

年代：清代
类别：古建筑
保护级别：青浦区文物保护单位
利用情况：开放参观
Era: Qing Dynasty
Category: Ancient architecture
Conservation level: Qingpu district-level protected site
Utilization: Open for visit

　　始建于元世祖至元年间（1264—1294），因由林姓老人出资所建，故名。又因桥北塌原有关帝阁，俗称"关爷桥"。清雍正八年（1730）重建。系单孔石拱桥，南北向，跨三里塘，桥长 21.3 米，宽 2.7 米，跨径 8.45 米，青石与花岗石材混砌，桥面两侧有低矮的石护栏、望柱和抱鼓石。

　　Linlao Bridge was originally built during the Zhiyuan era (Shizu's reign) of the Yuan Dynasty (1264–1294) and was named after an elderly man surnamed Lin who provided the funds for its construction. It is also known as "Guanye Bridge" due to the Guandi Pavilion at the northern end of the bridge. The bridge was reconstructed in the 8th year of the Yongzheng reign of the Qing Dynasty (1730). It is a single-arch stone bridge that spans the Sanli Pond in a north-south direction. The bridge is 21.3 meters long and 2.7 meters wide, with a span of 8.45 meters. It is built using a mixture of bluestone and granite, with low stone guardrails, wangzhu-balustrade pillars, and drum-shaped stones on both sides of the bridge deck.

如意桥
Ruyi Bridge

年代: 清代
类别: 古建筑
保护级别: 青浦区文物保护单位
利用情况: 开放参观
Era: Qing Dynasty
Category: Ancient architecture
Conservation level: Qingpu district-level protected site
Utilization: Open for visit

　　始建于元世祖至元年间,明崇祯年间(1628—1644)、清乾隆三十三年(1768)重修,光绪二十五年(1899)重建。系单孔石拱桥,南北向,跨东库港,桥长28.5米,宽2.96米,跨径8.6米,花岗石材质,桥面有护栏、望柱和抱鼓石,桥身外侧镌刻有"行道有福"题字。两侧桥柱上镌有楹联,东联为"后果前因如意桥发心遂意,顾名思义祖师庙主善为师",西联为"化险境为坦途,千秋如意;赖博施以济众,一路平安"。

　　Ruyi Bridge was originally built during the Zhiyuan era (Shizu's reign) of the Yuan Dynasty and underwent renovations in the Chongzhen reign of the Ming Dynasty (1628–1644), the 33rd year of the Qianlong reign (1768), and was rebuilt in the 25th year of the Guangxu reign of the Qing Dynasty (1899). It is a single-arch stone bridge that spans the Dongshe Harbor in a north-south direction. The bridge is 28.5 meters long and 2.96 meters wide, with a span of 8.6 meters. It is made of granite and has guardrails, wangzhu-balustrade pillars, and drum-shaped stones on both sides of the bridge deck. The outer side of the bridge is carved with the inscription "Blessings on the path". There are couplets engraved on the bridge pillars. The eastern couplet reads "The Ruyi Bridge embodies the will to achieve success, as the ancestral temple signifies the goodness of being a teacher". The western couplet reads "Transforming danger into a smooth path, eternal blessings; with generosity and assistance, a safe journey".

天皇阁桥
Tianhuangge Bridge

年代：清代
类别：古建筑
保护级别：青浦区文物保护单位
利用情况：开放参观
Era: Qing Dynasty
Category: Ancient architecture
Conservation level: Qingpu district-level protected site
Utilization: Open for visit

　　始建于明代，因桥北堍原有托塔天王庙，故名。清康熙三十七年（1698）重建。系三孔石拱桥，南北向，跨北胜浜，桥长21米，宽2.9米，并列分节拱券，中孔跨径6.8米，边孔跨径4米，青石与花岗石混砌，桥面雕有暗八仙图饰，东侧有楹联"愿天常生好人，愿人常行好事"。

　　Tianhuangge Bridge was originally built during the Ming Dynasty and was named after the Tuota Tianwang Temple at the northern end of the bridge. It was reconstructed in the 37th year of the Kangxi reign of the Qing Dynasty(1698). The bridge has three arches and spans the Beishengbang River in a north-south direction. It is 21 meters long and 2.9 meters wide, with segmented arches. The central arch has a span of 6.8 meters, while the side arches have a span of 4 meters. It is built using a mixture of bluestone and granite, and the bridge deck is carved with hidden Eight Immortals patterns. On the eastern side, there is a couplet that reads "May good people always be born, and may people always do good deeds".

金泽放生桥
Jinze Fangsheng Bridge

年代: 清代
类别: 古建筑
保护级别: 青浦区文物保护单位
利用情况: 开放参观
Era: Qing Dynasty
Category: Ancient architecture
Conservation level: Qingpu district-level protected site
Utilization: Open for visit

　　始建于明代，崇祯年间（1628—1644）重修，清乾隆五十六年（1791）重建，民国时加砌护栏及望柱。系单孔石拱桥，南北向，跨后浜，桥长25.8米，宽2.3米，拱径7.7米，花岗石材质。西侧刻楹联"水出湾潭通秀气，桥连如意接康衢"。

　　Originally built during the Ming Dynasty, Fangsheng Bridge was repaired during the reign of Chongzhen (1628–1644), and rebuilt in the 56th year of the Qianlong reign of the Qing Dynasty (1791). During the Republican era, railings and wangzhu-balustrade pillars were added. The bridge is a single-hole stone arch bridge, spanning the Houbang River in a north-south direction. It measures 25.8 meters in length, 2.3 meters in width, and has an arch diameter of 7.7 meters. It is made of granite. On the west side of the bridge, there is an engraved couplet "Water flows out of the pond connecting to the beautiful air, bridge links to happiness and fortune heading towards prosperous road".

颐浩寺遗址
Yihao Temple Ruins

年代: 宋代
类别: 古遗址
保护级别: 青浦区文物保护单位
利用情况: 宗教活动
Era: Song Dynasty
Category: Ancient ruins
Conservation level: Qingpu district-level protected site
Utilization: Religious activities

颐浩寺建于宋景定元年（1260），里人费辅之创建，相传为宰相吕颐浩舍宅为寺，就此得名。后又经几次扩展，元贞元年（1295）奉旨赐名"颐浩禅寺"，规模宏大，为当时著名的江南佛寺。《松江府志》称誉颐浩寺"虽杭之灵隐，苏之承天，莫匹其伟"。后历代多次修葺，1938 年大部分建筑毁于轰炸，现仅存元代"松江府颐浩寺记"碑 1 方、"不断云"石刻 1 处、大殿柱础 12 个、佛台基座 1 个和少量破损方砖。

The Yihao Temple was built in the 1st year of the Jingding reign of the Song Dynasty (1260) by Fei Fuzhi, a local resident. It is said to have been the residence of the prime minister Lü Yihao before it was turned into a temple, hence the name. It was expanded several times afterward, and in the 1st year of the Yuanzheng reign of the Yuan Dynasty (1295), it was granted the name "Yihao Zen Temple" by imperial decree. It was a large-scale temple and was famous in the Jiangnan region at that time. The "*Songjiang Prefecture Annals*" praised the Yihao Temple as "No match can be found for the grandeur of the Yihao Temple, even when compared to the Lingyin Temple in Hangzhou or the Chengtian Temple in Suzhou". It underwent several renovations throughout the dynasties but was mostly destroyed by bombing in 1938. Only a few elements remain, including a stele called "The Songjiang Prefecture Yihao Temple Record" from the Yuan Dynasty, a stone carving of "Continuous Clouds", twelve plinths of the main hall, one foundation of a Buddha's platform, and a small number of damaged square bricks.

川沙新镇

Chuansha New Town

1 黄炎培故居
川沙新镇新川路 218 号

2 宋氏家族居住纪念地
川沙新镇新川路 218 号

3 川沙古城墙
川沙新镇新川路 171 号旁 "川沙古城墙公园" 内

4 岳碑亭
川沙新镇新川路 171 号旁 "川沙古城墙公园" 内

5 川沙天主堂
川沙新镇中市街 42 弄 15 号

6 上川铁路川沙站旧址
川沙新镇华夏东路 2696 号

7 陶桂松住宅
川沙新镇操场街 48 号

8 丁家花园
川沙新镇北市街 19 号

1 *Huang Yanpei Residence*
No.218, Xinchuan Road, Chuansha New Town

2 *Memorial Site of Song Family Residence*
No.218, Xinchuan Road, Chuansha New Town

3 *Chuansha Ancient City Wall*
Inside "Chuansha Ancient City Wall Park", next to No.171 Xinchuan Road, Chuansha New Town

4 *Yue Stele Pavillion*
Inside "Chuansha Ancient City Wall Park", next to No.171 Xinchuan Road, Chuansha New Town

5 *Chuansha Catholic Church*
No.15, Lane 42, Zhongshi Street, Chuansha New Town

6 *Former Site of Chuansha Station of Shangchuan Railway*
No.2696, Huaxia East Road, Chuansha New Town

7 *Tao Guisong Residence*
No. 48, Caochang Street, Chuansha New Town

8 *Ding Family Garden*
No.19, Beishi Street, Chuansha New Town

图例 LEGENDS

🅝 全国重点文物保护单位
National priority protected site

🅝 上海市文物保护单位
Shanghai city-level protected site

🅝 区级文物保护单位
District-level protected site

🅝 区文物保护点
District-level protected place

◎ 其他景点
Other attractions

← 范围外景点
Out-of-range attractions

⊷⊷⊷ 历史风貌区范围
Scope of historical disctrict

游览路线
Touring route

华夏高架路 Huaxia Elevated Road

华夏高架路 Huaxia Elevated Road

东运盐河 Dongyunyan River

上川铁路川沙站旧址
⑥ Former Site of Chuansha Station
of Shangchuan Railway

车站路 Chezhan Road

东河浜路 Donghebang Road

城河浜 Chenghebang River

北城壕路 Beichenghao Road

北市路 Beishi Road

陶桂松住宅
⑦ Tao Guisong Residence

东河浜路 Donghebang Road

东运盐河 Dongyunyan River

浜路 Sanzaobang Road

西市街 Xishi Street

新川路 Xin Road

川沙天主堂
⑤ Chuansha Catholic Church

东门街 Dongmen Street

中市街 Zhongshi Street

川沙戏曲艺术展示中心
Chuansha Opera Art
Exhibition Center

乔家浜路 Qiaoliabang Road

石皮路 Shipi Road

新川路 Xinchuan Road

黄炎培故居
① Huang Yanpei Residence

宋氏家族居住纪念地
② Memorial Site of
Song Family Residence

川沙古城墙
Chuansha Ancient City Wall

丁家花园
⑧ Ding Family Garden

③

岳碑亭
④ Yue Stele Pavilion

川沙古城墙公园
Chuansha Ancient City Wall Park

城河浜 Chenghebang River

🅝 本图为位置示意，与实际尺寸不符
Illustration is not proportional to the actual scale

川沙新镇
Chuansha New Town

交通指南：
地铁 2 号线至川沙站。

Transportation Guide:
Metro Line 2 to Chuansha Station.

风水堡城，海派营造

川沙新镇位于浦东新区中部，长江入海口南岸，是因盐而兴、因商而聚、因纺而盛的工商名镇，曾长期是浦东地区的经济和文化中心，有着丰富的人文资源和历史积淀，素有"浦东历史文化之根"的美誉。

北宋时期，今川沙所在地逐渐成陆，成陆后不久海边便筑起捍海塘，塘内熬波晒盐，塘外水路运输，一时盐业兴盛。后陆域东移，原斥卤之地化为可垦良田，盐作文化演变为稻作文化。明代，上海有七大盐场，川沙地区隶属于其中的下砂三场，到万历年间，已是"生聚日繁，人文渐盛，巍然为滨海巨镇"。到了明代中叶，倭寇不断进犯，为防沿海倭患，川沙于明嘉靖三十六年（1557）修筑城堡，以此为中心大力发展轻纺手工业，繁荣之势辐射整个江南。清代，浦东制盐业走向衰落，不少人为解决生计，学当泥水工或木工，其中的一批能人巧匠逐步在上海开埠后的城市建设浪潮下成长为沪上知名的"营造厂商"，包括被誉为"浦东鲁班"一代宗师的

杨斯盛以及顾兰洲、赵增涛、陶桂松等。他们为开创"万国建筑博览群"的辉煌贡献了智慧和力量，成就了上海"远东第一大都市"的建筑根基，川沙也因此被誉为"营造之乡"。经济的繁荣带来的是川沙人文的兴盛。川沙自明代至今走出了各行各界的名人达100多位，如宋庆龄、黄炎培、林均、黄自、沈树镛等，为社会建设和文化发展作出了重要贡献。

川沙古城格局较为完整，至今保留有"口"字形护城河、"十"字形主街和部分古城墙。历史建筑数量众多，传统形式与中西合璧的乡土建筑交相辉映，沪剧、川沙民间故事、江南丝竹等非物质文化遗产依然延续，反映了浦东地区的建筑文化和原住民的生活形态。

黄炎培先生曾说"浦东文化在川沙"，川沙是浦东乃至上海现存城池格局较完整的古城建置聚落，同时，它也是先民滨海繁衍的历史写照，明代海防抗倭的真实遗存、众多名人的诞生地和近代民族工业发展的重要见证。围绕"风水堡城、海派营造、名家故里、戏曲之乡、宗教遗存"五个方面的特征，川沙正在全力打造兼具江南风情和文化内涵的文旅特色小镇，让游客能在度过悠闲、惬意的慢生活的同时感受传统文化的魅力。

川沙新镇参观指南 Chuansha New Town Visiting Guide

游览路线：
川沙戏曲艺术展示中心→丁家花园(川沙营造馆)→黄炎培故居（内史第）→川沙古城墙公园→川沙天主堂→上川铁路川沙站旧址

Tourist Route：
Chuansha Opera Art Exhibition Center → Chuansha Construction Museum) → Huang Yanpei Residence (Neishi House) → Chuansha Ancient City Wall Park → Chuansha Catholic Church → Former site of Chuansha Station of Shangchuan Railway

古镇美食特产：
塌饼、菜饭

Local Specialties：
Collapsed Cake, Vegetable Rice

不可移动文物资源

2014 年 2 月，川沙新镇被公布为第六批中国历史文化名镇，镇内划定中市街历史文化风貌保护区和六灶港历史文化风貌保护区。其中，中市街风貌区面积 82 公顷，核心保护区面积 12.1 公顷；六灶港风貌区面积 18.7 公顷，核心保护区面积 6.5 公顷。镇域内现有各级不可移动文物 53 处，包括上海市文物保护单位 1 处，区级文物保护单位 10 处，区文物保护点 42 处。建筑年代主要集中在清末与民国时期，砖木结构为主，具有江南传统建筑特色和中西合璧的装饰风格。

A town of fengshui fortress, famous for Shanghai-style architecture

Chuansha New Town is located in the central part of Pudong New Area, on the south bank of the mouth of the Yangtze River. It is a well-known industrial and commercial town that has prospered due to its salt production, gathered because of trade, and thrived thanks to its textile industry. It was once the economic and cultural center of the Pudong region and has rich cultural resources and historical heritage, earning the reputation as the "root of Pudong's history and culture".

During the Northern Song Dynasty, the area where Chuansha is located gradually emerged from the sea. Soon after becoming land, seawalls were built to protect against the sea, and salt was produced by evaporating seawater in the enclosed ponds while transportation was facilitated through waterways outside the walls. This led to a thriving salt industry. As the land shifted eastward over time, the areas previously used for salt production were transformed into cultivable farmland, and the culture shifted from salt production to rice farming. In the Ming Dynasty, there were seven salt fields in Shanghai, the Chuansha area belong to the Xiasha third (salt) field, and by the Wanli reign, it had become a thriving coastal town known for its abundant resources and flourishing culture. In the mid-Ming Dynasty, Chuansha faced constant attacks from Japanese pirates. In order to protect against these coastal threats, a castle was built in the 36th year of the Jiajing reign (1557). This castle served as the center for the development of the textile industry, which thrived and radiated throughout the entire Jiangnan region. During the Qing Dynasty, salt production in Pudong began to decline. Many people had to find alternative livelihoods and became skilled craftsmen in fields such as masonry or carpentry. Some of these talented craftsmen gradually grew into renowned construction companies in Shanghai during the city's development after its opening to the outside world. These included Yang Sisheng, hailed as the "Lu Ban of Pudong", as well as

Gu Lanzhou, Zhao Zengtao, Tao Guisong, and others. Together, they made contribution to the brilliant "buildings of variegated Western architectures", establishing the foundation for Shanghai to become the "largest metropolis in the Far East". Because of their contributions, Chuansha came to be known as the "hometown of construction". The economic prosperity of Chuansha has brought about a flourishing of its cultural scene. From the Ming Dynasty to the present, Chuansha has produced over 100 famous figures from various fields. These include Soong Ching-ling, Huang Yanpei, Lin Jun, Huang Zi, Shen Shuyong, and others. These individuals have made significant contributions to social development and cultural advancement.

Chuansha Ancient City retains a relatively complete layout, with a moat in the shape of the Chinese character " 口 ", a main street in the shape of the character " 十 ", and some remaining ancient city walls. There are numerous historical buildings, and the combination of traditional forms and the fusion of Chinese and Western architectural styles create a harmonious blend of local buildings. Intangible cultural heritage such as Shanghai opera, Chuansha folk stories, and Jiangnan silk and bamboo music continue to thrive, reflecting the architectural culture of Pudong and the lifestyle of the local inhabitants.

Mr. Huang Yanpei once said, "Pudong culture lies in Chuansha". Chuansha is one of the few remaining ancient cities in Pudong and even in Shanghai that retains a complete city layout. It is a historical portrait of the coastal settlements where the early inhabitants thrived. It is also a true relic of the defense against Japanese pirates in the Ming Dynasty, the birthplace of many famous figures, and an important witness to the development of modern national industry. With its distinctive features of "city of fengshui fortress, architecture of Shanghai style, hometown of renowned celebrities, hometown of traditional opera, relics of religious buildings", Chuansha is striving to become a cultural and tourism characteristic town that combines the charm of the Jiangnan region with cultural depth. Visitors can enjoy a leisurely and pleasant lifestyle while experiencing the allure of traditional culture.

Immovable cultural relics resources

In February 2014, Chuansha New Town was announced as the sixth batch of national famous historical and cultural towns. Within the town, the Zhongshi Street Historical and Cultural Conservation Area and Liuzao Port Historical and Cultural Conservation Area were designated. The Zhongshi Street Conservation Area covers an area of 82 hectares, with a core protected area of 12.1 hectares. The Liuzao Port Conservation Area covers an area of 18.7 hectares, with a core protected area of 6.5 hectares. There are currently 53 immovable cultural relics at all levels in the town, including 1 Shanghai city-level protected sites, 10 district-level protected sites and 42 district-level protected places. The buildings in the town mainly date back to the late Qing Dynasty and the Republic of China period, featuring traditional Jiangnan architecture with a fusion of Chinese and Western decorative styles, primarily constructed with brick and wood structure.

黄炎培故居
Huang Yanpei Residence

年代：清代
类型：近现代重要史迹及代表性建筑
级别：上海市文物保护单位
利用情况：开放参观
Era: Qing Dynasty
Category: Modern important historical sites and representative buildings
Conservation level: Shanghai city-level protected site
Utilization: Open for visit

黄炎培（1878—1965），中国教育家、社会活动家，中国近代职业教育的创始人和理论家。黄炎培故居是一座二层砖木结构楼房，坐北朝南，为"内史第"第三进，黄炎培在此出生。"内史第"为黄炎培舅舅沈树镛祖上所建，清咸丰九年（1859），沈树镛中举，官至内阁中书，遂改名"内史第"。故居前有仪门，为独立式八字墙门楼，尺度高大，飞檐翘角，中间有"华堂映日"石额，两侧石柱刻有"四福捧寿"纹，下端基石盘龙石刻。内有天井，粉墙黑瓦，飞檐翘壁，屋宇雕梁画栋，各类雕刻装饰尤为精致典雅。正楼前设有一座黄炎培半身铜像，上悬陈云同志手书"黄炎培故居"匾额。

Huang Yanpei (1878–1965) was a notable patriot and democratic educator, as well as the founder and theorist of modern vocational education in China. The Huang Yanpei Residence is a two-story brick and wood structure, where Huang Yanpei was born. It faces south and is the third section of the "Neishi House". The "Neishi House" was originally built by Huang Yanpei's uncle Shen Shuyong's ancestors. In the 9th year of the Xianfeng reign of the Qing Dynasty (1859), Shen Shuyong achieved success in the imperial examination and eventually became an official in the Imperial Cabinet, hence the name "Neishi House". In front of the residence, there is an ornate gateway in the form of an independent octagonal archway, with a tall scale and upturned eaves. In the middle, there is a stone plaque reading "Huatang Yingri" (meaning "The Splendid Hall Reflecting the Sunlight"), and the stone pillars on both sides are engraved with the patterns of "Four Blessings Holding Longevity". At the lower end, there is a carved dragon on the foundation stone. Inside, there is a courtyard with white walls and black tiles, with eaves and walls protruding. The building is adorned with exquisite carvings and decorations, showcasing elegant craftsmanship. There is a bronze statue of Huang Yanpei in front of the main building, with a plaque reading "Huang Yanpei Residence" written by Comrade Chen Yun.

宋氏家族居住纪念地

Memorial Site of Song Family Residence

年代：清代
类型：近现代重要史迹及代表性建筑
保护级别：浦东新区文物保护单位
利用情况：开放参观
Era: Qing Dynasty
Category: Modern important historical sites and representative buildings
Conservation level: Pudong New Area district-level protected site
Utilization: Open for visit

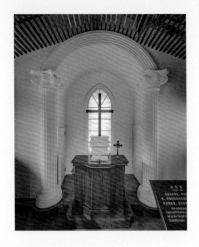

宋氏家族居住纪念地在"内史第"内。19世纪末，沈家将沿街前进出赁，清光绪十三年（1887），宋耀如与倪桂珍结婚，宋耀如、倪桂珍婚后赁屋于"内史第"西侧沿街房屋。

The residence of the Song family, a memorial site, is also located within the "Neishi House". In the late 19th century, the Shen family gradually expanded their properties along the street. In the 13th year of the Qing Dynasty's Guangxu reign (1887), after Song Yaoru married Ni Guizhen, the couple rented a house on the western side of the "Neishi House" along the street.

川沙古城墙
Chuansha Ancient City Wall

年代: 明代
类别: 古建筑
保护级别: 浦东新区文物保护单位
利用情况: 开放参观
Era: Ming Dynasty
Category: Ancient architecture
Conservation level: Pudong New Area district-level protected site
Utilization: Open for visit

　　明嘉靖三十六年（1557），为抵御倭寇入侵而建川沙堡城。建好后"城周四里，高二丈八尺，址阔三丈有奇，惶深如高之半，为门者四，楼如之"，清代重修。现尚存东城墙60余米，黄土夯筑，青砖包砌，是上海保存最长、最完整的明代城墙。2010年开辟为川沙古城墙公园。

　　In the 36th year of the Jiajing reign of the Ming Dynasty (1557), the Chuansha Fort was built to defend against Japanese pirates. After its completion, it had a circumference of four li, a height of two zhang and eight chi (1 chi is approximately 0.33 meters), and a width of three zhang. It had four gates and towers. The city wall was repaired during the Qing Dynasty. Presently, there are still over 60 meters of the eastern city wall remaining, constructed with rammed earth and covered in gray bricks. It is the longest and most intact Ming Dynasty city wall preserved within Shanghai. It was opened as the Chuansha Ancient City Wall Park in 2010.

岳碑亭
Yue Stele Pavilion

年代： 清代
类别： 石窟寺及石刻
保护级别： 浦东新区文物保护单位
利用情况： 开放参观
Era: Qing Dynasty
Category: Grottoes, temples and stone carvings
Conservation level: Pudong New Area district-level protected site
Utilization: Open for visit

清道光十二年（1832），川沙同知郑其忠在种德寺摹勒宋朝名将岳飞赠李梦龙手书诗于石，1913年移入观澜书院（今观澜小学），1929年移至城墙上，并建亭，取名"岳碑亭"。石碑为青石质，方首，碑上诗为："学士高僧醉如泥，玉山颓倒瓮头低。酒杯不是功名具，入手缘何只自迷。"后附题记："商丘狂学士李梦龙索余书□□大梁之舞剑阁，岳飞草。"后亭毁，1987年重建碑亭，为二层方形歇山顶楼阁式建筑。后再次重建，恢复为民国时期风貌。

In the 12th year of the Qing Dynasty's Emperor Daoguang (1832), a local official named Zheng Qizhong from Chuansha inscribed and engraved the hand-written poem by the Song Dynasty general Yue Fei presented to Li Menglong on a stone at Zhongde Temple. In 1913, the stele was moved to Guanlan Academy (now Guanlan Primary School). In 1929, it was relocated to the city wall and a pavilion was built to house it, named Yue Stele Pavilion. The stele is made of green stone, with a square shape and a poem inscribed on it. The poem reads, "Scholars and monks are drunk like mud, the jade mountain collapsed with a tilted gourd. The wine glass is not a vessel for fame and success, so why am I lost in it?" It is followed by an inscription, "The unrestrained scholar Li Menglong from Shangqiu requested me to write a poem... Sword Dance Pavilion of Da Liang, Yue Fei". The pavilion was later destroyed, but in 1987 it was rebuilt as a two-story, square-shaped building with a gable and hip roof. It was then rebuilt again to restore its appearance from the Republican era.

川沙天主堂
Chuansha Catholic Church

年代: 近现代
类型: 近现代重要史迹及代表性建筑
保护级别: 浦东新区文物保护单位
利用情况: 宗教活动
Era: Modern times
Category: Modern important historical sites and representative buildings
Conservation level: Pudong New Area district-level protected site
Utilization: Religious activities

川沙天主堂又称耶稣圣心堂,始建于清同治十一年(1872)。后由海门郁兰生(神甫黄重裳的亲戚)出资改建大堂,于民国十五年(1926)竣工。建筑外形是哥特式单钟塔,平面呈拉丁十字式,大厅为巴西利卡式。钟楼位于主立面入口上部,顶端为尖锥形塔尖,内悬铜钟3只。外墙为青、红两色相间砌成的清水墙,门窗均为尖券。教堂内部为典型的哥特式建筑风格,正坛供奉耶稣圣心像,左右分供圣母、若瑟怀抱耶稣像各一尊。

The Chuansha Catholic Church, also known as the Sacred Heart of Jesus Church, was originally built in the 11th year of the Qing Dynasty's Tongzhi reign (1872). It was later renovated and expanded by Yu Lansheng from Haimen (a relative of Father Huang Chongchang), and completed in the 15th year of the Republic of China (1926). The architectural style of the church is Gothic, with a single bell tower and a Latin cross-shaped floor plan. The hall is in the style of a basilica. The bell tower is located above the main entrance and has a pointed spire at the top, with three bronze bells hanging inside. The exterior walls are brick walls without plastering. The doors and windows are all pointed arches. The interior of the church showcases typical Gothic architectural style. The main altar is dedicated to the Sacred Heart of Jesus, and there are statues of the Virgin Mary and Joseph holding the baby Jesus on either side.

上川铁路川沙站旧址

Former Site of Chuansha Station of Shangchuan Railway

年代： 近现代
类型： 近现代重要史迹及代表性建筑
保护级别： 浦东新区文物保护单位
利用情况： 开放参观
Era: Modern times
Category: Modern important historical sites and representative buildings
Conservation level: Pudong New Area district-level protected site
Utilization: Open for visit

　　清末民初，浦东地区交通不便，1921 年黄炎培等人筹建上川铁路，1925 年上川铁路由庆宁寺连接至川沙。上川铁路是上海第一条商办轻轨铁路，它的建成极大地便利了川沙乃至南汇地区人们出行交通。1975 年拆除，后按原址复制了"川沙火车站旧址"，位于川沙烈士陵园门口，站台为青砖铺设，围有木栅栏，并保存着 20 世纪 30 年代蒸汽机车头。

In the late Qing Dynasty and early Republic of China, transportation in the Pudong area was inconvenient. In 1921, Huang Yanpei and others initiated the construction of the Shangchuan Railway. In 1925, the Shangchuan Railway was connected to Chuansha from Qingning Temple. The Shangchuan Railway was the first commercial light rail in Shanghai and greatly facilitated the transportation of people in Chuansha and even Nanhui region. It was demolished in 1975, and a replica of the "Former site of Chuansha Train Station" was built on the original site. It is located at the entrance of the Chuansha Martyrs Cemetery. The platform is paved with green bricks and surrounded by wooden fences. It also preserves a steam locomotive from the 1930s.

陶桂松住宅
Tao Guisong Residence

年代： 近现代
类型： 近现代重要史迹及代表性建筑
保护级别： 浦东新区文物保护单位
利用情况： 居住场所
Era: Modern times
Category: Modern important historical sites and representative buildings
Conservation level: Pudong New Area district-level protected site
Utilization: Residence

陶桂松住宅系上海著名的营造商陶桂松为自己修建的宅邸，又名"陶氏精舍"，建于1930年。建筑分为主楼、副楼两部分。主楼二层，坐北朝南，为一正两厢布局，砖木混合结构，天井上方有玻璃雨棚，主入口设中式仪门，朝内为砖石质牌科门楼，地面为彩色拼花马赛克地砖等，门窗上有彩色玻璃。副楼位于东侧，为砖混结构二层楼房，平屋顶。整体呈现中式布局与西式装饰结合，装饰精美、体量宏大，是川沙古镇近代民居建筑的精品之一。

The Tao Guisong Residence, also known as the "Tao Family Residence", is a residential building in Shanghai built by the renowned builder Tao Guisong in 1930. The building is divided into a main building and an auxiliary building. The main building is two stories high, facing south, with a symmetrical layout of one main hall and two wing rooms. It is constructed using a combination of bricks and wood, with a glass canopy above the courtyard. The main entrance features a traditional Chinese ceremonial gate, while the interior features a brick and stone plaque gatehouse. The ground is decorated with colored mosaic tiles and the windows have stained glass. The auxiliary building is located on the east side, a two-story brick and concrete structure with a flat roof. The overall design combines Chinese layout with Western-style decorations, showcasing exquisite craftsmanship and a grand scale. It is one of the masterpieces of modern urban residential architecture in Chuansha Ancient Town.

丁家花园
Ding Family Garden

年代: 近现代
类型: 近现代重要史迹及代表性建筑
保护级别: 浦东新区文物保护点
利用情况: 开放参观

Era: Modern times
Category: Modern important historical sites and representative buildings
Conservation level: Pudong New Area district-level protected place
Utilization: Open for visit

丁家花园，建于 1935 年，店主丁云石（桃生）建造。平面布局为三合院形式，二层砖木结构建筑，占地 560 平方米。门前有一个花园式庭院，故得名丁家花园。建筑的梁架结构形式为穿斗式木结构，东南朝向。正厅高二层，与厢房共同形成面向院落的三面围合的廊道。屋顶形式为硬山顶，皆为小青瓦双坡屋顶。1949 年后为川沙县第一任政府所在地，后为县级机关大楼，现作为川沙营造馆使用。

Ding Family Garden, built in 1935, was constructed by the owner Ding Yunshi (Taosheng). The layout of the garden follows the form of a semi-closed courtyard, with a two-story brick and wood structure covering an area of 560 square meters. There is a garden-style courtyard in front, hence the name Ding Family Garden. The building's beam and frame structure are of a column and tie construction wooden structure and faces southeast. The main hall is two stories high and, together with the wing rooms, forms a three-sided corridor facing the courtyard. The double-sloped flush gable roof is all covered with small gray tiles. After 1949, it served as the first government office of Chuansha County, and later as a building for county-level institutions. It is now used as Chuansha Construction Museum.

罗店镇   Luodian Town

1 梵王宫
       罗店镇罗溪路 518 号

1 *Fanwang Palace*
       No.518, Luoxi Road, Luodian Town

2 花神堂
       罗店镇赵巷街 136 号

2 *Huashen (Flower Deity) Hall*
       No.136, Zhaoxiang Street, Luodian Town

3 罗店红十字纪念碑
       罗店镇罗太路 382 号（陈伯吹中学内）

3 *Luodian Red Cross Monument*
       Inside Chen Bochui Middle School, No.382,
       Luotai Road, Luodian Town

4 钱世桢墓
       罗店镇毛家弄村三树南路以东，圃南路以南

4 *Qian Shizhen's Tomb*
       East of Sanshu South Road, South of Punan
       Road, Maojianong Village, Luodian Town

5 远景村王宅
       罗店镇远景村东小路西小塘子 2 号

5 *Yuanjing Village Wang Residence*
       No.2, West Xiaotangzi, Dongxiao Road, Yuanjing
       Village, Luodian Town

6 敦友堂
       罗店镇亭前街 284 号；亭前街 288 弄 5 号

6 *Dunyou Hall*
       No.284 Tianqian Street; No.5, Lane 288,
       Tingqian Street, Luodian Town

7 大通桥
       罗店镇新桥居委亭前街弄口

7 *Datong Bridge*
       Xinqiao Community, Tingqian Street, Luodian
       Town

8 来龙桥
       罗店镇市一路 130 号罗溪公园内

8 *Lailong Bridge*
       Inside the Luoxi Park, No.130, Shiyi Road,
       Luodian Town

图例 LEGENDS

- **n** 全国重点文物保护单位
  National priority protected site
- **n** 上海市文物保护单位
  Shanghai city-level protected site
- **n** 区级文物保护单位
  District-level protected site
- **n** 区文物保护点
  District-level protected place
- ◎ 其他景点
  Other attractions
- ← 范围外景点
  Out-of-range attractions
- ▪▪▪ 历史风貌区范围
  Scope of historical disctrict
- ━━ 游览路线
  Touring route

祁北路 Qjbei Road

练泾 Dijing

来龙桥
Lailong Bridge
8

4

扬新河 Liangj River

宝山寺
Baoshan Temple

① 梵王宫
Fanwang Palace

钱世桢墓
Qian Shizhen Tomb

② 花神堂
Huashen (Flower Deity) Hall

市一路 Shiyi Road

东西巷街 Dongxixiang Street

市沙河 Shiho an Road

花神广场
Huashen Square

塘桥街 Tangxi Street

⑥ 敦友堂
Dunyou Hall

亭前街 Tingqian Street

⑦ 大通桥
Datong Bridge

③ 罗店红十字纪念碑
Luodian Red Cross Monument

练祁河 Shiha River

东西巷街 Dongxixiang Street

罗溪路 Liuoxi Road

蕴川公路 Uechuan Road

练泾 Dijing

沪太路 Old Qijing
Highway

⑤ 远景村王宅
Yuanjing Village Wang Residence

🧭 本图为位置示意，与实际尺寸不符
Illustration is not proportional to the actual scale

罗店镇
Luodian Town

交通指南:
地铁 7 号线至罗南新村站，步行 1000 米可达。

Transportation Guide:
Metro Line 7 to Luonan Xincun Station,
1000 meters walk.

江口集镇，面海而生

上海周边的市镇，清代以来曾经流传一句耳熟能详的顺口溜："金罗店、银南翔、铜江湾、铁大场。""罗店"被誉为"金"，排名第一。

罗店镇位于宝山区西北部，是上海北翼的重要门户。早在1400多年前，这一地区冲积成陆，唐开元元年（713）筑捍海塘，已有人烟。12世纪后，浏河港兴起，毗邻的罗店地区水上航运逐步兴盛。元至正年间（1341—1368），商人罗升在此设店，形成集市，得名"罗店"，又因位于练祁河北，又名"罗阳""罗溪"。

随着江南沿海棉业及贸易的兴旺，罗店凭借通江达海的水运交通，成为苏浙皖棉业交汇之地，明代前期已成大镇，清康熙年间成为棉花、棉布交易贸易中心。清末民初，该镇形成了"三湾九街十八弄"的集镇规模，三里长街有花行、米行、布庄、酱园等商铺六七百家，每日三市，市面繁荣，为全县最大的市镇，遂有"金罗店"之称。清人范连曾作《罗溪杂咏》：

"练水西来清且涟，波光近与界泾边。不须更访罗昇宅，烟火今经五百年。"直至近代，罗店棉业仍是上海轻纺工业发展的组成部分。

除了商业的发展，罗店地处要冲，屏藩上海，北邻郑和下西洋的出发地浏河港，东南拱卫着东海要塞吴淞口，是沪北海防的重要防线，也是抗击外侮的要塞。自明代开始，罗店人一直有保家卫国的不屈节操。明嘉靖年间，抗击倭寇。清顺治二年（1645），抵抗清兵。近代两次淞沪抗战，罗店都是主要战场，"八一三淞沪战役"中更为战略要地，老镇建筑几近瓦砾残垣。

虽然近代几经兵燹，但罗店古镇格局和街巷肌理相对完整，仍具有史称的"街衢综错，宛如棋枰绮脉之形"。镇内商铺和住宅毗邻，街巷和河道交织，依稀可辨当年商贸重镇的风采。其中保存较好的亭前街，街两边老式店面毗连，保存有清代建筑"敦友堂""承恩堂"等。

罗店非物质文化遗产资源丰富，现有国家级非物质文化遗产，端午节罗店划龙船习俗，上海市非物质文化遗产罗店龙船、罗店彩灯、传统木结构建筑营造技艺（宝山寺唐式木结构营造技艺），区级非物质文化遗产罗店鱼圆、天花玉露霜、公大酱制品工艺、罗店版画、罗店民俗画、痔漏传统治疗法等。每年端午节，罗店都会举办"龙船文化节"，延续划龙船习俗，在美兰湖上进行精彩纷呈的龙船表演，同时还会组织特色的民俗行街，以及制作龙船模型、彩灯、鱼圆、天花玉露霜、手工香囊、面塑、十字挑花、竹编等非遗活动。在罗店，"春有花神秋有画，夏有龙船冬有灯"，一年四时，都能感受不同的民俗文化。

遗韵犹存的水乡空间，类型多样的建筑遗产，独树一帜的非遗传承，崇文重教的优良传统，保家卫国的斗争精神，赋予罗店独特的历史文化基因，也体现了江南文化在江口沿海区域既交汇融合又外向创新的长江口商贸集镇特色。

不可移动文物资源

2019 年 1 月，罗店被公布为第七批中国历史文化名镇，其中历史文化风貌保护区面积 80.95 公顷，核心保护区面积 12.2 公顷。镇域内共有不可移动文物 25 处，包括上海市文物保护单位 1 处，区级文物保护单位 4 处，区文物保护点 20 处。

罗店镇参观指南 Luodian Town Visiting Guide

游览路线：
梵王宫（宝山寺）→花神堂→来龙桥→花神广场→大通桥→敦友堂→远景村王宅

Tourist Route：
Fanwang Palace (Baoshan Temple) → Huashen (Flower Deity) Hall → Lailong Bridge → Huashen Square → Datong Bridge → Dunyou Hall → Yuanjing Village Wang Residence

古镇美食特产：
罗店鱼圆、天花玉露霜、民俗画、古镇汤圆、草头饼

Local Specialties：
Luodian Fish Balls, Tianhua Yulu Cream, Folk Painting, Ancient Town Dumplings, Caotou Cake

A town by the river, born facing the sea

In the towns around Shanghai, there has been a well-known traditional rhyme since the Qing Dynasty that goes: "Gold Luodian, Silver Nanxiang, Copper Jiangwan, Iron Dachang." Among these towns, "Luodian" is called Gold and is ranked first in importance.

Located in the northwest part of Baoshan District, Lodian Town is an important gateway to the north of Shanghai. More than 1400 years ago, this area was formed by alluvial deposits. In the 1st year of the Tang Dynasty's Kaiyuan reign (713), seawall protection was built here to defend against the sea, and there was already human habitation. In the 12th century, Liuhe Port flourished, and the adjacent Luodian area gradually prospered in water transportation.

During the Yuan Dynasty's Zhiyuan period (1341–1368), the merchant Luo Sheng set up a shop here and formed a market, which was named "Luodian". Because it is located north of the Lianqi River, it is also known as "Luoyang" or "Luoxi".

With the prosperity of the cotton industry and trade along the south coast of Jiangnan, Luodian, with its water transportation connecting rivers and seas, became a gathering place for the cotton industries of Jiangsu, Zhejiang, and Anhui provinces. In the early Ming Dynasty, it had become a major town, and during the Kangxi period of the Qing Dynasty, it became a trading center for cotton and cotton fabrics. In the late Qing Dynasty and early Republic of China, the town developed into a market town with the scale of "Three Bays, Nine Streets, and

Eighteen Alleys". The three-li-long street had six to seven hundred shops, including florists, rice shops, cloth stores, and sauce and pickle shops. The market was prosperous with everyday three markets, making it the largest market town in the county, and it was known as "Gold Luodian". In the late Qing Dynasty, Fan Lian, a poet, wrote in his poem "Miscellaneous Verses of Luoxi": "The Lian River comes from the west, clear and flowing; The waves are close to the Jie River. No need to visit Luo Sheng Residence, the smoke and fire have been here for five hundred years." Until modern times, the cotton

industry in Luodian remained an integral part of the development of Shanghai's textile industry.

In addition to its commercial development, Luodian is strategically located as a key defense line in the north of Shanghai. It is located near Liuhe Port, the starting point of Zheng He's voyages to the West, and it guards the Wu Song mouth, an important fortress in the East China Sea. It has been an important stronghold in Shanghai's coastal defense and a fortress against foreign aggression. From the Ming Dynasty onwards, the people of Luodian have always had a strong sense of patriotism. During the

Jiajing period of the Ming Dynasty, they fought against Japanese pirates. In the 2nd year of the Shunzhi reign (1645), they resisted the Qing army. During the modern period, Luodian was a major battlefield in the two Battles of Shanghai during the War of Resistance Against Japanese Aggression, and it played a crucial role in the "August 13th Battle of Shanghai", leaving the old town in ruins.

Although there have been several wars and destruction in modern times, the layout and street texture of Luodian ancient town remain relatively intact. It still retains the historical name "streets and lanes interwoven like a chessboard and exquisite blood vessels". The town is filled with shops and residences, with streets and waterways intertwining, faintly revealing the charm of the former commercial town. Among them, the well-preserved Tingqian Street has a continuous row of old-fashioned storefronts, and it also retains Qing Dynasty buildings such as "Dunyu Hall" and "Cheng'en Hall".

Luo Dian possesses rich intangible cultural heritage resources. There is one national-level intangible cultural heritage item, which is the Dragon Boating Customs of Luodian during the Dragon Boat Festival. There are three municipal-level intangible cultural heritage items in Shanghai, namely Luodian Dragon Boat, Luodian Lanterns, and Traditional Wood Structure Construction Skill (Tang-style Wood Structure Construction Skill of Baoshan Temple). Additionally, there are district-level intangible cultural heritage items including Luodian Fish Balls, Tianhua Yulu Cream, Gongda Jiang (a traditional soybean paste) Fermented Products Craftsmanship, Luodian Block Printing, Luodian Folk Paintings, and Traditional Treatment

Methods for Hemorrhoids, etc. Every year during the Dragon Boat Festival, Luodian holds the "Dragon Boat Culture Festival" to continue the Dragon Boating Customs. There will be splendid dragon boat performances on the Meilan Lake, as well as organized folk parades featuring making dragon boat models, lanterns, fish balls, Tianhua Yulu cream, handmade sachets, dough art, cross-stitched embroidery, bamboo weaving, and other intangible cultural activities. In Luodian, "there are flower gods in spring, paintings in autumn, dragon boats in summer, and lanterns in winter". Throughout the year, you can experience different folk cultures.

The water town of Luodian showcases a preserved cultural landscape, diverse architectural heritage, unique folk beliefs, distinctive intangible cultural heritage, a rich tradition of cultural cultivation, and a spirit of patriotism and defense. These elements have endowed Luodian with a unique historical and cultural heritage, reflecting the characteristics of the Yangtze River Delta's commercial and trade town where Jiangnan culture converges, integrates, and innovates in the coastal region along the Yangtze River estuary.

Immovable Cultural Relics Resources

In January 2019, Luodian was announced as the seventh batch of national famous historical and cultural towns. The historical and cultural landscape protection area covers an area of 80.95 hectares, with a core protection area of 12.2 hectares. There are a total of 25 immovable cultural relics in the town, including 1 Shanghai city-level protected sites, 4 district-level protected sites and 20 district-level protected places.

梵王宫
Fanwang Palace

年代：清代
类别：古建筑
保护级别：宝山区文物保护单位
利用情况：宗教活动
Era: Qing Dynasty
Category: Ancient architecture
Conservation level: Baoshan district-level protected site
Utilization: Religious activities

梵王宫始建于明正德（1506—1521）、嘉靖年间（1522—1566），原为唐氏住宅，后改为佛寺。清乾隆二十七年（1762）增建真武阁。光绪五年（1879）重建，前为王灵宫殿，中为玉皇宫，后为玄帝阁。现仅存大雄宝殿及祖堂塔院，即现在的天王殿。大雄宝殿为二层砖木结构，面阔五间，穿斗式梁架，歇山顶。天王殿为一层砖木结构，面阔三间，歇山顶，殿内左侧竖立石碑1方及弥勒石佛像1尊。梵王宫的历次改建、易名，呈现出由私家宅院到佛寺、真武阁，再到大雄宝殿的独特历程，实际体现了清代江南底层民众"佛道合抱"的民间信仰状况，罗店地区特有的祭祀玉皇的正月十九庙会，曾经以此为主要场所。

The Fanwang Palace, also known as Tang Residence, is believed to have been originally built during the Ming Dynasty's Zhengde period (1506–1521) and further expanded during the Jiajing period (1522–1566). It was originally a residence of the Tang family but later converted into a Buddhist temple. In the 27th year of the Qianlong period of the Qing Dynasty (1762), the Zhenwu Pavilion was added. In the fifth year of the Guangxu period (1879), the palace was rebuilt, with the front part being the Wang Ling Palace, the middle part being the Yu Huang Palace, and the rear part being the Xuan Di Palace. Only the Mahavira Hall and the Zutang Ta Yuan (Ancestral Hall Tower Courtyard), now known as the Tian Wang Hall, still exist. The Mahavira Hall is a two-story brick and wood structure with a width of five bays, a gable and hip roof and column and tie construction. The Tian Wang Hall is a one-story brick and wood structure with a width of three bays and a gable and hip roof. Inside the hall, there is a stele and a statue of Maitreya Buddha on the left side. The successive renovations and renaming of the Fawang Palace reflect its unique transformation from a private residence to a Buddhist temple, Zhenwu Pavilion, then to the Mahavira Hall. It reflects the popular belief in Buddhism and Taoism among the lower-class people in the Jiangnan region during the Qing Dynasty. The temple hosted the main temple fair for the worship of the Yu Huang (Jade Emperor) on the 19th day of the lunar month, which was a unique event in the Luodian area.

花神堂
Huashen (Flower Deity) Hall

年代: 清代
类别: 古建筑
保护级别: 宝山区文物保护点
利用情况: 其他用途
Era: Qing Dynasty
Category: Ancient architecture
Conservation level: Baoshan district-level protected place
Utilization: Other uses

始建于明天启年间（1621—1627），原为城隍行宫建筑的一部分。清咸丰十一年（1861）毁于战火，清光绪十三年（1887）重建。建筑坐北朝南，硬山顶式，哺鸡式屋脊，面阔 11.48 米，进深 7.8 米，占地 89.5 平方米。殿内扁作鹤胫轩，荷包梁、轩梁及步柱上饰以浮雕花纹，廊下有葵式挂落，殿屋前院门有砖雕门楼，门楣南向字碑处题有行书"花业公所"砖雕字样，为当地书法家朱六阶所书。北向字碑处有行书"万花主宰"砖雕字样，出自晚清状元洪钧之手。花神堂反映出罗店地区特有的棉花崇拜和花神信仰，也是罗店棉业兴衰的见证。

The Huashen (Flower Deity) Hall was originally built during the Tianqi period of the Ming Dynasty (1621–1627) and was part of the Chenghuang Temple complex. It was destroyed during the war in the 11th year of the Xianfeng period of the Qing Dynasty (1861) and was rebuilt in the 13th year of the Guangxu period (1887). The building faces south and has a flush gable roof with chicken-broth-shaped ridges. It has a width of 11.48 meters, a depth of 7.8 meters, and covers an area of 89.5 square meters. Inside the hall, there are carved beams and rafters with floral patterns, and the corridors are decorated with hanging ornaments in the shape of sunflowers. The front gate of the hall has a gatehouse with brick carving, and on the lintel of the gate, there is a calligraphy inscription in running script that reads "Hua Ye Gong Suo" (The Flower Industry Guild) which was written by local calligrapher Zhu Liujie. On the north side of the gate, there is another calligraphy inscription in running script that reads "Wan Hua Zhu Zai" (The Master of Ten Thousand Flowers) which was created by Hong Jun, a top scholar in the late Qing Dynasty. The Huashen Hall reflects the unique cotton worship and flower deity belief in the Luodian area and serves as a witness to the rise and fall of the cotton industry in Luodian.

罗店红十字纪念碑
Luodian Red Cross Monument

年代： 近现代
类别： 近现代重要史迹及代表性建筑
保护级别： 上海市文物保护单位
利用情况： 开放参观
Era: Modern times
Category: Modern important historical sites and
representative buildings
Conservation level: Shanghai city-level protected site
Utilization: Open for visit

　　始建于民国三十五年（1946），为纪念 1937 年 8 月 23 日因救治和掩护对日空战中负伤的飞行员苑金函而遭侵华日军杀害的中国红十字会上海分会第一救护队副队长苏克己和队员谢惠贤、刘中武、陈秀芳而建。1981 年 6 月，旧碑年久失修，当地民政局于罗店中学内重立。1984 年市文管委又在原碑西侧 15 米处按原样放大重建纪念碑。现碑高 8.3 米，正面题字刻隶书阴文"中华民国红十字会总会第一救护队抗战殉难烈士纪念碑"字样，四周以阴文刻四烈士遇难经过，东西两侧镶烈士瓷像。

The Luodian Red Cross Monument was built in 1946 to commemorate the Chinese Red Cross Society Shanghai Branch's First Aid Team Deputy Captain Su Keji and team members Xie Huixian, Liu Zhongwu, and Chen Xiufang. They were killed by the invading Japanese army on August 23, 1937, while providing medical treatment and protection to wounded pilots Yuan Jinhan during the air battle with the Japanese. On June 1981, local bureau of Civil administration rebuilt a monument inside Luodian Middle School for its bad condition. The monument was enlarged and rebuilt in its original style in 1984, 15 meters west of the original one by the Shanghai Municipal Bureau of Cultural Affairs. The current monument is 8.3 meters high, and the front of the monument is inscribed with the words "Memorial Monument to the Martyrs of the First Rescue Team of the Federation of the Red Cross Societies of the Republic of China in Anti-Japanese War" in clerical script. The four martyrs' martyrdom experience are engraved in yin script on four sides, while porcelain painting of four martyrs are embedded on east and west sides.

钱世桢墓
Qian Shizhen Tomb

年代: 明代
类别: 古墓葬
保护级别: 宝山区文物保护单位
利用情况: 开放参观
Era: Ming Dynasty
Category: Ancient burial
Conservation level: Baoshan District-level protected site
Utilization: Open for visit

建于明崇祯年间（1628—1644）。墓主钱世桢，明万历十七年（1589）武进士，历任金山参将、蓟辽总兵等职，著有《征东实记》，为明代抗倭名将。其墓南向，有祭台和甬道，原两侧分列翁仲石马，并植有银杏多棵。

Qian Shizhen Tomb was built during the Chongzhen period of the Ming Dynasty (1628–1644). Qian Shizhen, the owner of the tomb, passed the martial examination in the 17th year of the Wanli reign (1589) and held various military positions such as the general in Jinshan and the commander in charge of Jizhou and Liaodong. He is known for his book "*The Record of Conquering the East*" and is a famous general of the Ming Dynasty who fought against Japanese pirates. The tomb faces south and is equipped with a sacrificial platform and a passage. Originally, stone horses and ceremonial guards were placed on both sides, and several ginkgo trees were planted.

远景村王宅

Yuanjing Village Wang Residence

年代: 清代
类别: 古建筑
保护级别: 宝山区文物保护点
利用情况: 开放参观
Era: Qing Dynasty
Category: Ancient architecture
Conservation level: Baoshan district-level protected place
Utilization: Open for visit

　　始建于清代，俗称"西小塘子塘南村宅"，因房主姓王，故称王宅。宅院面阔25.3米，进深27.4米，占地面积693平方米，为一层砖木结构传统四合院式民宅。宅院东西两侧各有厢房多间，北侧为大小不同中式房。宅院梁柱等整体结构和风格保存较为完整。现作为罗店书画院使用。

Yuanjing Village Wang Residence, also known as "West Xiaotangzi Tangnan Village Residence" in the vernacular, was originally built during the Qing Dynasty. It is commonly referred to as Wang Residence because the owner's surname is Wang. The residence covers an area of 25.3 meters in width, 27.4 meters in depth, and occupies a total area of 693 square meters. It is a one-story brick and wood structure traditional courtyard-style residence. There are multiple wing rooms on both the east and west sides of the residence, and on the northern side is a Chinese-style room of varying sizes. The overall architectural style and structure of the residence, including the beams and pillars, have been well-preserved. It is currently used as the Luodian Painting and Calligraphy Academy.

敦友堂
Dunyou Hall

年代: 清代
类别: 古建筑
保护级别: 宝山区文物保护点
利用情况: 其他用途
Era: Qing Dynasty
Category: Ancient architecture
Conservation level: Baoshan district-level protected place
Utilization: Other uses

始建于清代，为清朝孝子朱渐义（1652—1738）后人建造。现留存有厅堂一间、门楼两座。厅堂坐北朝南，通面阔 10 米。大院通进深 82 米，为一层砖木结构，建筑占地面积 828 平方米，南侧房门东沿留存雌毛脊五山屏风墙一堵。坐南朝北，有牌科，门楼字碑处题有"承家堂构"行书阴文。另有"居仁行义"门楼，坐南朝北，字碑处题刻"居仁行义"隶书阳文。

Dunyou Hall, originally built during the Qing Dynasty, was constructed by the descendants of Zhu Jianyi (1652–1738), a filial son of the Qing Dynasty. Currently, it still retains one hall and two gatehouses. The hall faces south and has a width of 10 meters. The courtyard extends 82 meters in depth and is a one-story brick and wood structure. The building occupies an area of 828 square meters. On the east side of the southern entrance, there is a preserved wall with five mountain-shaped firewalls with Cimao ridge. Next to it stands a gatehouse called "Cheng Jia Tang Gou" facing north with a plaque, which is inscribed with the words "Cheng Jia Tang Gou"(a place where the family property is inherited or taken over) in seal script and is written in shadow script. There is also another gatehouse called "Ju Ren Xing Yi" practicing benevolence and righteousness facing north with the words "Ju Ren Xing Yi" (practicing benevolence and righteousness) inscribed in clerical script on the plaque.

大通桥
Datong Bridge

年代：清代
类别：古建筑
保护级别：宝山区文物保护单位
利用情况：开放参观
Era: Qing Dynasty
Category: Ancient architecture
Conservation level: Baoshan district-level protected site
Utilization: Open for visit

　　始建于明成化八年（1472），清雍正八年（1730）重建，清道光二十八年（1848）、2005 年重修，跨罗店市河，南北向，单孔石拱桥，长 15.79 米，宽 4.52 米，高 4.50 米，拱径 7.86 米。桥额题刻正书阳文桥名，另在望柱上题刻"道光二十八年正月"和"里人重修"正书阳文。桥西侧拱洞两旁题刻有桥联"前程路途通万里，津梁岁月亘千秋"，因年代久远，字迹已模糊。

　　Datong Bridge, originally built in the 8th year of the Ming Dynasty's Chenghua reign (1472), rebuilt in the 8th year of the Qing Dynasty's Yongzheng reign (1730), repaired in the 28th year of the Qing Dynasty's Daoguang reign (1848) and in 2005. It spans the Luodian River in a north-south direction and is a single-arch stone bridge. It is 15.79 meters long, 4.52 meters wide, 4.50 meters high, with an arch diameter of 7.86 meters. The bridge is inscribed with the bridge name in regular script on the bridge plaque, and on the wangzhu-balustrade pillar it is inscribed with "January of the twenty-eighth year of Daoguang" and "Repaired by the local residents" in regular script. On the west side of the bridge arch, there is a couplet inscription that reads, "The road ahead connects thousands of li, the bridge stands for millions of years". Due to its age, the inscriptions have become blurred.

来龙桥
Lailong Bridge

年代: 清代
类别: 古建筑
保护级别: 宝山区文物保护点
利用情况: 开放参观
Era: Qing Dynasty
Category: Ancient architecture
Conservation level: Baoshan district-level protected place
Utilization: Open for visit

　　建于清同治八年（1869），原位于宝山区罗店镇蒋家巷街北，为南北向、跨老练祁河的单孔石拱桥。1992 年被迁移至罗溪公园内异地保护。现为东西向单孔石拱桥。桥长 12.2 米，宽 2.97 米，高 4.72 米，拱径 5.84 米。桥额题刻正书阴文"来龙桥"桥名，桥两端均为石阶式桥塂，栏杆上雕有石刻小狮子。桥两侧各镌有正书阴文桥联"彩虹晴锁练江潮，此来彩笔海题桥""紫气晓回沧海归，此去银涛皆入海"。

　　The Lailong Bridge was built in the 8th year of the Qing Dynasty's Tongzhi reign (1869), originally located in Jiangjia Lane, Luodian Town, Baoshan District. It is a single-arch stone bridge spanning the Lianqi River in a north-south direction. In 1992, it was relocated to the Luoxi Park for preservation. Now it is an east-west single-arch stone bridge. The bridge is 12.2 meters long, 2.97 meters wide, and 4.72 meters high. The arch has a diameter of 5.84 meters. The name of the bridge, "Lailong Bridge" is inscribed in calligraphy on the bridge face. Both ends of the bridge are stone steps, and the railings are carved with stone lions. There are two sets of inscriptions in calligraphy on both sides of the bridge: "Rainbow locks the Lianjiang tide on a clear day, and this colorful bridge is painted by the sea" and "Purple air returns to the sea at dawn, and the silver waves all enter the sea on this journey".

后记

　　江南文化底蕴悠长、内涵丰富，既是中华优秀传统文化的重要组成部分，也是长三角区域共同的文化标识，更是上海城市文化的重要根源。

　　2018年起，上海市文化和旅游局（上海市文物局）着力推进"建筑可阅读"工作，深受欢迎。为推进"建筑可阅读"深度发展，打响上海文化品牌，弘扬城市文脉，上海市文物保护研究中心推出了建筑可阅读书系之文物视角中的江南系列，尝试从不可移动文物的角度阐释江南文化的丰富内涵。古镇作为江南文化集大成者，蕴含江南文化基因，饱含江南水乡韵味，是一个整体和动态的江南文化缩影，因此将《文物视角中的江南：上海古镇》作为本书系的首部推出。

　　成书之际，感谢各级领导的关怀，感谢各专家学者的协助，感谢各区文物管理部门、各镇人民政府、文体中心、旅游公司的大力支持，感谢同济大学出版社的严谨刊布以及本书的责任编辑、美术编辑的辛勤付出。期望本书的出版，能带动更多读者关注上海古镇，探寻古镇中的优秀文化遗产。

　　由于水平有限，时间仓促，难免有不足之处，敬请批评指正。

编者

2023年8月31日

Epilogue

The cultural heritage of Jiangnan is long-standing and rich in content. It is not only an important part of China's outstanding traditional culture but also a common cultural symbol in the Yangtze River Delta region. Moreover, it serves as an important source of Shanghai's urban culture.

Since 2018, the Shanghai Municipal Administration of Culture and Tourism (Shanghai Municipal Administration of Cultural Heritage) has been actively promoting the "Stories of Shanghai Architecture" work, which has been well received. In order to further develop the "Stories of Shanghai Architecture" initiative, establish the cultural brand of Shanghai, and promote the urban context, the Shanghai Cultural Heritage Conservation and Research Center has launched "Jiangnan from the perspective of Cultural Relics" volumes of "Stories of Shanghai Architecture" series, attempting to interpret the rich connotations of Jiangnan culture from the perspective of immovable cultural relics. As the epitome of Jiangnan culture and the carrier of its cultural genes, ancient towns are imbued with the charm of Jiangnan's water towns. They represent a holistic and dynamic portrayal of Jiangnan culture. Therefore, *Jiangnan from the Perspective of Cultural Relics: Ancient Towns in Shanghai* has been selected as the inaugural volume of this series.

On the occasion of the publication, we would like to express our gratitude to the leaders at all levels for their attention and support. We sincerely appreciate the support from experts and scholars. We would also like to extend our thanks to the cultural relics management departments, people's governments of various towns, cultural and sports centers, and tourism companies in each district for their generous support. We are grateful to Tongji University Press for their meticulous publication and to the responsible editor and art editor for their hard work. We hope that the publication of this book will encourage more readers to pay attention to the ancient towns of Shanghai and explore the outstanding cultural heritage within these towns.

Due to our limited expertise and time constraints, there may be some shortcomings. We welcome and appreciate any criticism and suggestions from everyone.

Editor

August 31, 2023

枫泾镇 Fengjing Town

嘉定镇 Jiading Town

金泽镇 Jinze Town

枫泾镇 Fengjing Town

图书在版编目（CIP）数据

文物视角中的江南：上海古镇：汉英对照 / 上海
市文物保护研究中心编著 . -- 上海：同济大学出版社，
2024.3
（建筑可阅读书系）
ISBN 978-7-5765-0996-0

Ⅰ . ①文… Ⅱ . ①上… Ⅲ . ①乡镇－古建筑－建筑艺
术－上海－汉、英 Ⅳ . ① TU-092.2

中国国家版本馆 CIP 数据核字 (2024) 第 037179 号

本书中的部分研究获国家社会科学基金资助（项目号：16BGL186）

文物视角中的江南

上海古镇

上海市文物保护研究中心　编著

出 品 人　金英伟
责任编辑　由爱华　金　言
责任校对　徐春莲
书籍设计　张　微
出版发行　同济大学出版社 www.tongjipress.com.cn
　　　　　（地址：上海市四平路 1239 号　邮编：200092　电话：021‑65985622）
经　　销　全国新华书店
印　　刷　上海雅昌艺术印刷有限公司
开　　本　787mm×1092mm　1/36
印　　张　7
字　　数　206 000
版　　次　2024 年 3 月第 1 版
印　　次　2024 年 3 月第 1 次印刷
书　　号　ISBN 978-7-5765-0996-0
定　　价　68.00 元